AF564560

Food and Industrial Bioprocessing

NIPA® GENX ELECTRONIC RESOURCES & SOLUTIONS P. LTD.
New Delhi-110 034

Food and Industrial Bioprocessing

Pinki Saini
Neelam Yadav
Centre of Food Technology
University of Allahabad
Prayagraj, Uttar Pradesh

NIPA® GENX ELECTRONIC RESOURCES & SOLUTIONS P. LTD.
New Delhi-110 034

NIPA® GENX ELECTRONIC
RESOURCES & SOLUTIONS P. LTD.
101,103, Vikas Surya Plaza, CU Block
L.S.C.Market, Pitam Pura, New Delhi-110 034
Ph : +91 11 27341616, 27341717, 27341718
E-mail:newindiapublishingagency@gmail.com
www: www.nipabooks.com

For customer assistance, please contact
Phone: + 91-11-27 34 17 17
Fax: + 91-11- 27 34 16 16

ISBN: 978-81-19002-53-5

NIPA® also publishes books in a variety of electronic formats. Some content that appears in print may not be available in electronic books, and vice versa

Composed and Designed by NIPA®.

Preface

Food bioprocessing is a green technology that offers an alternative means of food and food ingredient extraction, purification and production using processes that involves the application of enzymes and/or microorganisms. It is one of the earliest forms of food modifying method used by early Egyptians in the production of wine, beer, and bread. Studies on the properties, production and applications of food enzymes and microorganisms have led to innovative and value-added products to be marketed such as low-lactose milk, probiotics dairy produce, clarified fruit juices, bioactive peptide-rich food and beverages, and many more. A relatively new concept in enzyme application is in situ modulation or transformation of the physico-chemical and sensory properties, and stability of local agricultural and food produce and commodities including fats and oils. In the same light, food microorganisms can be used to modify the properties of food through fermentation, serve as starter cultures for better quality and consistent food products, and to produce industrial food ingredients. The adoption of bioprocessing can bring about the opening of new channels for a wider usage of local commodities especially those that are underutilized and for better consumer acceptance since bioprocessing is ecofriendly. The existing industrial, government and research & development infrastructure in biotechnology alongwith the pervasive influence of biological substances in everyday life, has set the stage for unprecedented growth of bioprocessing sector in products, markets and expectations.

The objective of the book is to compile a detailed review of various developments in the area of food and industrial bioprocessing in a form that is easily comprehensible across a broad spectrum of readers working in relevant essential disciplines. The book reviews the development and implementation of processes for the manufacture of antibiotics and other pharmaceuticals, pigments, biocatalysts, industrial sugars, bioenergy, alcohols, amino acids, organic acids, nanocomposites packaging materials and specialty products through the application of microbiology, fermentation, enzymes and separation technology. Reports on future prospects of capability and application of bioprocessing for the benefit of society have also been addressed. The book is envisaged as a complete material that could be applied as a teaching text, reference material for advanced learners and source book for researchers, practitioners and technicians in research laboratories in the relevant disciplines. We thank all the contributors to this book for sharing their expertise in the subject. We hope that readers find the book content rewarding and enriching.

Pinki Saini
Neelam Yadav

Contents

List of Contributors

A Sujatha, Department of Biochemistry, Biotechnology and Bioin-formatics, Avinashilingam Institute for Home Science and Higher Education for Women, Coimbatore-641043 Tamil Nadu, India

A. Shobana, Department of Biochemistry, Biotechnology and Bioinformatics, Avinashilingam Institute for Home Science and Higher Education for Women, Coimbatore-641043 Tamil Nadu

Aishwarya Vetrivel, Department of Biochemistry, Biotechnology and Bioinformatics Avinashilingam Institute for Home Science and Higher Education for Women Coimbatore, Tamil Nadu, India

Amrita Poonia, Department of Dairy Science and Food Technology, Institute of Agricultural Sciences, Banaras Hindu University, Varanasi -221005, Uttar Pradesh

Anchal Singh, Centre of Food Technology, University of Allahabad, Prayagraj - 211002 Uttar Pradesh

Anitha Subash, Department of Biochemistry, Biotechnology and Bioinformatics, Avinashilingam Institute for Home Science and Higher Education for Women Coimbatore-641043 Tamil Nadu

Anusha Ramani, Centre of Food Technology, University of Allahabad, Prayagraj - 211002 Uttar Pradesh

Devinder Kaur, Centre of Food Technology, University of Allahabad, Prayagraj - 211002 Uttar Pradesh

Ena Gupta, Department of Family and Community Science University of Allahabad Prayagraj – 211002, Uttar Pradesh

M.Priyadarshini, Department of Biochemistry, Biotechnology and Bioin-formatics Avinashilingam Institute for Home Science and Higher Education for Women Coimbatore-641043, Tamil Nadu

Madhulika Bhagat, School of Biotechnology, University of Jammu Jammu and Kashmir, 180006

Mazia Ahmed, Centre of Food Technology, University of Allahabad, Prayagraj - 211002 Uttar Pradesh

Neelam Dubey, Centre of Food Technology, University of Allahabad, Prayagraj - 211002 Uttar Pradesh

Neelam Yadav, Centre of Food Technology, University of Allahabad, Prayagraj - 211002 Uttar Pradesh

Pinki Saini, Centre of Food Technology, University of Allahabad, Prayagraj - 211002 Uttar Pradesh

Poonam Yadav, Department of Food Science and Technology, College of Agriculture, G.B. Pant University of Agriculture and Technology, Pantnagar - 263145 Uttarakhand

Pragya Mishra, Department of Food Processing and Management, DDU Kaushal Kendra RGSC, BHU, Varanasi Uttar Pradesh

R R Mishra, Medical Laboratory Technology, DDU Kaushal Kendra, RGSC, Banaras Hindu University, Vara-nasi-221005, Uttar Pradesh, India

R.Shruthi, Department of Biochemistry, Biotechnology and Bioin-formatics, Avinashilingam Institute for Home Science and Higher Education for Women,Coimbatore-641043 Tamil Nadu, India

Rafiq A Rather, School of Biotechnology, University of Jammu, Jammu and Kashmir, 180006

Rajeswari Murugesan, Department of Biochemistry, Biotechnology and Bioinformatics

Shanthy Sundaram, Centre of Biotechnology, Nehru Science Centre, University of Allahabad, Prayagraj – 211002, Uttar Pradesh

Shashi Soni, Department of Family and Community Science, University of Allahabad Prayagraj, Uttar Pradesh, India, 211002

Sonia Thapa, Cancer Pharmacology Division, CSIR-Indian Institute of Integrative Medicine Canal Road, Jammu, 180001

Suman Devi, College of Dairy Technology, Etawah Campus, Chandra Shekhar Azad University of Agriculture & Technology, Kanpur – 208002, Uttar Pradesh

Swati Gupta, Centre of Biotechnology, Nehru Science Centre, University of Allahabad Prayagraj – 211002, Uttar Pradesh

Unaiza Iqbal, Centre of Food Technology, University of Allahabad, Prayagraj – 211002 Uttar Pradesh

Uroosa Noor, 1Department of Family and Community Science, University of Allahabad Prayagraj, Uttar Pradesh, India, 211002

Urvashi Srivastava, Centre of Food Technology, University of Allahabad Prayagraj - 211002, Uttar Pradesh

Vinita Puranik, Centre of Food Technology, University of Allahabad, Prayagraj-211002 Uttar Pradesh

1

Prospects and Issues of Bioprocessing

Neelam Yadav and Anusha Ramani

Centre of Food Technology, University of Allahabad, Prayagraj Uttar Pradesh

Abstract

Bioprocessing is much greener than chemical methods, since biocatalysts (enzymes or cells) involved are extremely selective, producing higher product yield with less or no by products, which are basically difficult to separate. Biotechnology can offer both economic and environmental benefits to the chemical industry and thus has great potential to achieve the sustainable production of existing and new products from renewable feedstocks. Fuel and energy production from biomass are major market sector for bioprocessing. They are considered in the manufacturing of biopharmaceuticals, food, food additives, chemicals, and several other products. Generally, a bioprocess involves feedstock pretreatment, fermentation or biocatalysis, and downstream processing for obtaining product and its purification. It is explored for the production of xylitol, fatty alcohol, butanol and biorefinery. Deep understanding of the process is required to create robust control strategy. Based on the requirements, sophisticated instrumentation and complex control laws or simple control law having intricate monitoring system can be selected. Supercritical fluids (SCFs) are advantageous for bioprocessing due to its transport properties, high diffusivity and low viscosity facilitate mass transfer phenomena, variation of pressure and/or temperature are adjusted providing high flexibility in the extraction of metabolites from cells. Microfluidic devices in which small amounts of fluids (10^{-9} and 10^{-18}) are manipulated have emerged for high throughput applications. Automation of the optimization processes is enabled through microtiter plates with simplicity and operate with very small volumes. Therefore, more research is necessary to develop integrated processes that will allow combining one or more techniques to enhance bioprocessing at lower costs.

Introduction

High concern about environmental pollution and the sustainability of natural resources have paved way for swift transformation of chemical and pharmaceutical manufacturing industries, to get converted to more environmentally benign and bio-based. Bioprocessing history dates back to 10,000-7000 BC, when Dionysus developed wine in Eastern Mediterranean. Since then, the progress continued in this field through development of bread, cheese, fermentation proved by Pasteur, vinegar, yoghurt, penicillin, DNA structure, recombinant DNA, human insulin, amino acids, xylitol, succinates etc (Yang et al., 2007).

The production of chemicals through biotechnological approach is much greener than chemical methods, since biocatalysts (enzymes or cells) involved are extremely selective, producing higher product yield with less or no by products, which are basically difficult to separate. Biotechnology can offer both economic and environmental benefits to the chemical industry and thus has great potential to achieve the sustainable production of existing and new products from renewable feedstocks. Fuel and energy production from biomass are major market sector for bioprocessing. Biofuels, including ethanol, biodiesel, butanol, methane, and hydrogen made from renewable resources such as sugarcane and corn are replacing fossil fuels (Hess, 2006). Generally, a bioprocess involves feedstock pretreatment, fermentation or biocatalysis, and downstream processing for obtaining product and its purification (Figure 1).

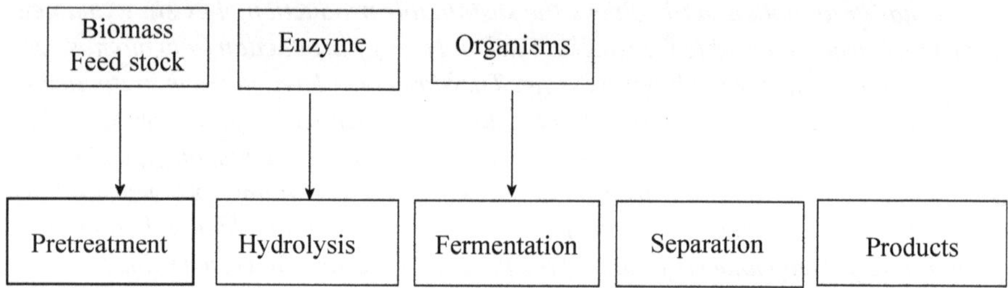

Fig. 1: A general bioprocess flow diagram

All types of microorganisms are used in industrial fermentation processes including bacteria, yeasts, and filamentous fungi. The choice for a fermentation process is mainly dependent upon the product, available substrate, and growth requirements, which govern the fermentation design and downstream processing. Nowadays, genome breeding, genome engineering, and genome shuffling permit effective whole-cell engineering of industrial strains leading to high-throughput screening techniques favourable for the discovery of new biocatalysts, organisms, and biologically active compounds at a faster rate (Petri and Schmidt-Dannert, 2004). Mainly submerged or solid state fermentations occurs, with former dominating. Mixing, heat and mass transfer inside large-scale reactor vessels are governed by bioreactor engineering. Cell immobilization provides a viable solution in improving productivity.

Downstream processing governed by several separation techniques based on differences in their size, solubility, density, volatility and partition coefficient in two different phases (Basu and Yang, 2005). Downstream processing generally comprises cell (particle) separation by filtration, sedimentation or centrifugation. Then it is followed by primary separation or enrichment by evaporation, precipitation or adsorption and secondary separation or purification done by distillation, crystallization. Then finally product is dried, polished and packed in case of powders. For liquid sterilization and particle separation, membrane processing mainly microfiltration is done, whereas for concentrating macromolecules such as proteins and polysaccharides, ultrafiltration is used and for removing inorganic salts, reverse osmosis and diafiltration is employed. Electrodialysis is also important to separate on charge difference.

The main biomass feed stock used in bioreactor, includes any organic matter that is available on a renewable or recurring basis. By 2030, The U.S. Department of Energy and the Department of Agriculture has projected that over 1 billion dry tons of biomass feedstock annually to produce 5% of power (heat and electricity), 20% of liquid transportation fuels (ethanol and biodiesels), and 25% of industrial products (chemicals and materials) (Oak Ridge National Laboratory, 2005). Main fermentation feedstocks used in bioreactor industry starch and sugar from agricultural crops such as corn, as they are easier to hydrolyse to glucose by either chemical or enzymatic methods than cellulose. Enzymes used for starch conversion are α-Amylase, β-Amylase, glucoamylase, pullulanase (Yang et al., 2007). The other major biomass which is most abundantly found in practically all plant-derived materials, ranging from wood, grass to agricultural residues and municipal solid wastes are lignocelluloses which are mainly cellulose, hemicellulose, and lignin. Cellulose, major organic component of plant (50%) is made up of D-glucose linked by β-glucosidic bonds. Complex heteropolysaccharides are the hemicelluloses comprising D-xylose, D-glucose, D-mannose, D-galactose, L-arabinose, D-glucuronic and D-galacturonic acids. Lignin occurs in plant cell wall is an amorphous heteropolymer of phenylpropane units, mainly derived from coniferyl alcohol, sinayl alcohol, and coumaryl alcohol. The enzymes employed for their hydrolysis are cellulases - endoglucanases (endo-1,4-β-glucanases, EG), cellobiohydrolases (exo-1,4-β-glucanases, CBH), and cellobiase (β-glucosidase) (Perez et al., 2003).

Bioreactor

Culture systems to produce cells or organisms are defined into bioreactors and fermenters. They are considered in various applications, like basic research and development, alongwith the manufacturing of biopharmaceuticals, food, food additives, chemicals, and several other products. Cells such as mammalian cell lines, insect cells, and stem cells, microorganisms (like bacteria, yeasts, and fungi), as well as plant cells and algae can be cultivated in bioreactors and fermenters. When bacteria, yeast, or fungi is cultivated then often the term used is fermenter. Bioreactor is often related to the cultivation of mammalian cells, but is also generically used (Camire et al., 2007)

Several types of bioreactors exist, but mainly stirred-tank bioreactors are preferred. Cultivation takes place in the bioreactor tank and the culture is mixed by stirring (instead of shaking). These bioreactors have different sizes (for cultures of a few milliliters to thousands of liters) and are made of various materials (usually glass, plastic or stainless steel). The basic components and functioning of stirred-tank bioreactors are always the same. A stirred-tank bioreactor system comprises several parts. A vessel that is filled with the medium in which cells are cultivated and is closed by head plate in which components attached to measure and adjust the culturing conditions, such as feed lines, sensors and control software. In a bioreactor, the culture is stirred with an impeller. The temperature of the culture medium is continuously monitored with a temperature sensor and regulated by placing it in a thermowell, wrapped with a heating blanket or has a water jacket. Required oxygen is transferred from the surrounding air to the culture medium. With the help of spargers, the gas/liquid interface can be increased and the oxygen supply maximized. Oxygen is important for culture growth, and the amount of oxygen dissolved (DO) in the medium is continuously measured with a DO sensor. In bioreactors, the medium pH is continuously measured using a pH sensor and CO2 is added as needed. In bioreactors, cultures are usually maintained for hours, days, or weeks, depending on the organism and application. (Suttle and Sha, 2018).

A bioprocess run typically comprises the following steps

1. **Preculture:** The medium in the bioreactor is inoculated with a preculture which may be grown in a shaker or incubator. Sometimes, smaller bioreactors are used to grow precultures.

2. **Bioreactor preparation:** Parallel to the inoculum preparation, the bioreactor is prepared by the sterilization of bioreactor, feed lines, and sensors; medium addition to the bioreactor; connection of the bioreactor with the bioprocess control station; and the setting of process parameter control points in the bioprocess control software.

3. **Inoculation:** After the bioreactor is prepared, the medium is inoculated.

4. **Cultivation period:** During the cultivation period in bioreactor, agitation, pH, temperature, and DO are typically monitored and controlled in real time via the bioprocess control software. In addition, culture samples are taken for analysis, such as biomass and the concentration of metabolites and eventually, nutrient solutions are added. Cultures typically pass through four growth phases, namely lag phase, exponential growth phase, stationary phase followed by the death phase.
5. **Culture harvest:** Typically, at the end of the bioprocess run, the culture is harvested when it enters the stationary growth phase.
6. **Downstream processing:** The culture broth is further processed.
7. **Bioreactor cleaning:** Lastly, the bioreactor is sterilized to inactivate culture residues and cleaned.

Bioreactors help in resolving cultivation bottlenecks as they may save work, time, and lab-space of scientists who need large quantities of cells, microbes, or of the products they express. Furthermore, it also improves reproducibility of growth, product formation, and product quality and increase cultivation efficiency. Small molecules generated by microbes and intended for use as chemical building blocks, fuels or food and feed additives may be required in small quantities in the R&D phase, but mostly in large quantities are needed for their commercial use. Stirred-tank bioreactors are scalable, so that the work and space requirements stay manageable (Dorceus, 2018).

Prospects of Bioprocessing

Some of the major prospects of bioprocessing are discussed as follows-

Biorefinery through bioprocessing

Edible or non-edible food that is rejected in any step of the food chain is called as food waste (FW). In primary sector manufacturing steps, FW is composed of food that do not reach consumer, as it is either discarded or recycled. In logistics (during transportation) and storing operations comprise non-edible food pieces, foods not complying with organoleptic standards, food from process failures or batches with microbial contamination, expiration date label, quality losses. Post-consumer steps include residues from household and food service sector (mainly restaurants and hospitals). Generally, this FW is made up of peels, bones, cooking oil, discharged food portions, etc (Carmona-Cabello et al., 2018). FW is non- hazardous, except animal waste and is suitable for incineration if the lower heating value is above 3344 kJ/kg (Xu et al., 2016a). But, green-house gas emissions (GHG) increase (Bernstad and Jansen, 2012). Hence, biorefinery is environmentally friendly recovery approach (Chatzifragkou et al., 2014; Dorado et al., 2009). Biorefinery is based on biomass, and may produce electricity, fuels, combined heat and power, analogous to petroleum refinery. It will help in reducing energy costs and GHG. Therefore, FW is a solid candidate as biorefinery raw material. FW nutrients are soluble in water to an extent of 74%, hence water activity (aw) range from 0.5 to 0.6 is adequate for fungi and yeast growth, while higher values (0.8-0.9) are required for bacteria (Ruijter et al., 2004).

In FW, carbohydrates represent the main carbon contribution for the production of hydrogen, bioethanol (Kumar et al., 1998), lactase (Shirai et al., 2001) or bio-oil (Maina et al., 2017) through the metabolism of microorganisms during fermentation. Even the fiber (hemicellulose, cellulose and pectin) are available in sufficient quantities to be used on a large scale. Compartmentalization or emulsion formation can retain nutrients, making necessary the use of pretreatments (Galanakis, 2015). Fibres and lignins restrict polysaccharide breakdown due to intricate structure. Delignification is a crucial parameter for economic viability of biorefineries. Hence, an adequate pretreatment should focus on increasing the accessibility of microorganisms. Physical methods of pretreatment (milling, freezing, extrusion, microwave or pyrolysis) are costly and are focused on increasing the surface area and accessibility to microorganisms. Alkaline chemical pretreatments may harm environment. Therefore, biological treatments are environmentally friendly and may use fungi, *Phanerochaete chrysosporium* or microbial consortium. Combined pretreatments including, enzyme hydrolysis together with superfine grinding and steam explosion is considered a top methodology with balanced energy consumption and optimized technical parameters (Procentese et al., 2018). Nitrogen for bioprocesses is provided by hydrolysis of protein molecules producing peptides and amino acids essential for microorganism growth (Gänzle et al., 2007). Proper C/N ratio is required which is affected by lipids also. Low C/N ratios and nitrogen degradation in FW (due to acidogenesis during fermentation) causing rapid acidification of medium, ultimately inhibit fermentation (Sajjadi et al., 2016). Lipids are heterogeneous group of molecules present in FW, basically composed of triglycerides (glycerol bonded to fatty acid chains), phospholipids, prostaglandins or terpenes, steroids (Barik and Paul, 2017). They are metabolised by β-oxidation, which provides energy to cells, thus producing intermediate metabolites. Yeast may use lipids from the initial substrate composition to produce ex novo lipid accumulation (Papanikolaou & Aggelis, 2011). They produce bioenergy (i.e. bioethanol, biodiesel or biohydrogen), high value-added bio-products (i.e. bio-oil, wax, bioplastics, lactic acid, tri- and di-carboxylic acids, biosurfactants or biolubricants) with applications in pharmaceutical and food industries. Biorefinery involves solid-state fermentation (SSF), where polysaccharides and proteins are split into monosaccharides and amino acids, respectively. Submerged fermentations are based on biological enzymatic hydrolysis. In FW, there could be metal and other contaminants which may affect biological processes (mostly enzymatic activity) or oxidative-reductive potential. Heavy metals such as Cd, Pb or Zn, may inhibit amylase, cellulase or urease synthesis (Nelson et al., 2008; Tornberg, 2005). Inhibition reduces nutrient transport and membrane function (Ewert et al., 2018). Iron is important for hydrogen fermentative production, as it is essential in the oxidation of pyruvate to acetyl-CoA (Elbeshbishy et al., 2017). Monovalent and divalent ions (K, Mg and Ca) increase activation as seen in a-amylase during ethanol fermentation by yeast (Sajjadi et al., 2016).

Xylitol bioproduction

Xylitol is a polyol, sweetening power similar to the conventional sugar but with lower caloric content, which is why used in food industry. It is metabolized independently of insulin, hence, a safe product for diabetic people (Mussatto, 2012). It is also valorised in feedstock integrated biorefineries, where through catalysis, it produces 1,2-propandiol, ethylene glycol, glycerol, lactic acid (Xu et al., 2020). In 2020, its global market reached US$ 921 Million (Research & Markets Reports, 2021). In market, it is currently chemically synthesized through catalytic hydrogenation of xylose and requires high-cost catalyst and severe reaction conditions with high energy expenditure due to the high temperature and pressure. Due to further purification, xylitol is more expensive product, as compared to other sugars. Therefore, biotechnological production has emerged as low-cost, eco-friendly alternative and with energetic benefits, achieved by saving energy during the production process or by the coproduction of energy in a biorefinery. This approach could be future reality, by employing adaptive evolution, genetic engineering of microorganisms, and techno-economic studies of the entire conversion chain (upstream, fermentation and downstream steps) (Queiroz et al., 2022).

In chemical process, catalytic hydrogenation of hydrolysed lignocellulosic biomass, followed by the xylose purification from the hemicellulosic fraction is done (Rao et al., 2016; Arcaño et al., 2020). Non-sugar compounds, other monomeric sugars are to be removed for recovery of pure xylitol. An ultrapure xylose solution is transferred to catalytic hydrogenation reaction, at high temperature and pressure, using Raney nickel as catalyst delivering up to 98% conversion yield. Then catalyst is removed and hydrogenated solution is concentrated using activated carbon adsorption and ion exchange chromatography for decolorization and inorganic salt removal, respectively (Gallezot, 2013).

In biotechnological route, mainly microbial conversion the most often used is enzymes. Main procedure include pretreatment of lignocellulosic biomass, concentration and detoxification of hemicellulosic hydrolysate, bioconversion of xylose to xylitol and its purification and recovery (Hernández-Pérez et al., 2009). To breakdown the recalcitrant structure of the lignocellulose and release xylose in the first step, non-selective treatments mainly acid hydrolysis, alkaline hydrolysis, water-based treatments, or mechanical disruption, organosolv are employed (Mussatto, 2016). Acid addition is disadvantageous, because of corrosion problems in the reactor, hence, hydrothermal (or liquid hot water (LHW)) pretreatment is effective method and more environmentally friendly alternative (Carvalheiro et al., 2016). Not only corrosion, but acid treatment releases arabinose, glucose, and acetic acid from hemicellulosic, causing the formation of furfural and 5-hydroxymethylfurfural due to the dehydration of pentoses and hexoses. Cellulose and lignin after deconstruction release glucose and several phenolic compounds, respectively. These non-sugar compounds inhibit microbial metabolism and its performance during the fermentation. Therefore, detoxification is done by activated charcoal (Marton et al., 2006) or in ion exchange resins (Santos et al., 2014), nanofiltration and reverse osmosis membranes (Nguyen et al., 2015), use of enzymes such as lignin peroxidases (Cho et al., 2009), biopolymers (Silva-Fernandes et al., 2017). Yeasts are used for conversion, as they able to achieve high yields at different scales of production. Generally, *Candida* (mainly *C. guilliermondii, C. parapsilosis,* and *C. tropicalis*), as well as the yeasts *Scheffersomyces stipitis, Debaryomyces hansenii*, and *Kluyveromyces marxianus* are employed. The main enzyme in yeast is xylose reductase (EC 1.1.1.21, XR), which converts xylose-to-xylitol is dependent on reduced cofactor NADH/NADPH. Enzyme xylitol dehydrogenase NAD^+-dependent (EC 1.1.1.9, XDH) can convert accumulated xylitol to xylulose which gets phosphorylated to xylulose-5-phosphate and converted into glycolytic intermediates, such as glyceraldehyde-3-phosphate (GA3P) and fructose-6-phosphate (F6P), through the phosphate pentose pathway (PPP). Therefore, further conversion is inhibited by the expansion of genetic engineering tools and strain development, for the production of an efficient industrial process for biotechnological production of xylitol. The downstream process of xylitol purification and recovery from the fermentation broth, separating residual sugars, phenolic compounds, salts, proteins require traditional methods of ion exchange resins, activated charcoal, chromatography, liquid-liquid extraction, and nanofiltration giving 40–60% recovery yield with the final product of 98% purity (Misra et al., 2011; Kumar et al., 2019; Faneer et al., 2017). Supercritical CO_2 used for xylitol recovery has recovery yield of 40.5%, but xylitol was recovered with 99.5% purity, which makes it promising alternative (Silva et al., 2020).

Amongst the various strategies for strain improvement are adaptive laboratory evolution (ALE) of microorganisms as it potentially improves the microbial performance during the fermentation by making cells adaptive with particular traits, to the specific stress conditions used such as medium containing lignocellulosic inhibitors (furan derivatives, organic acids and phenolic compounds) (Koppram et al., 2012). Using strain tolerance, hydrolysate detoxification step can be avoided. ALE process can be done by sequential transfer of the microbial culture to a new medium containing enlarged concentration of inhibitory compounds, or by using a chemostat system. One such process utilizing *Saccharomyces cerevisiae* in a chemostat system lead to a strain with higher xylitol yield (0.74 g/g) than with the wild-type strain (0.61 g/g) (Koppram et al., 2012). In ALE of a recombinant xylose using *K. marxianus* NIRE-K1 evolved in medium containing 2% of xylose as sole carbon source, had xylose consumption (more than 80%) and xylitol yield was 1.65 times higher compared to the non-adapted yeast (Sharma et al., 2016). Researchers explored yeast *S. stipitis* CBS 6054, which expressed *Xyl1* gene, encoding for enzyme xylose reductase, as a result 95% efficiency of xylose conversion into xylitol was achieved. This gene was identified in other species also namely *C. tropicalis* (Kim et al., 2019) and *K. marxianus* (Park et al., 2019). Engineered strain *K. marxianus* KCTC17555 reported xylitol productivity and yield of 1.00 g/L/h and 0.75 g/g, respectively in a 2.5-L bioreactor (Park et al., 2019). In *S. cerevisiae*, identification of the hexose transporter genes (*Hxt*) is highly regulated depending on concentration and availability of glucose to the yeast and it modulates other carbon source consumption (Elbing et al., 2004). Another strategy to improve the xylose accumulation by yeasts is co-substrate fermentation such as glucose, studied for NADPH regeneration (Ko et al., 2006). However, their ratio is to be considered. Several different glucose: xylose ratios: 1:25, 1:12, 1:5, and 1:2.5, were tested throughout the cultivation of *C. guilliermondii* in sugarcane bagasse hemicellulosic hydrolysate and was reported that 1:5 was found as being the best condition for xylitol assimilation (Silva et al., 2006). The cause of this effect is transport system mechanism and its higher affinity for glucose than xylose. As in yeast this system is controlled according to the available carbon source and the cell's sensitivity towards their concentration. Protein present on plasma membrane of *S. cerevisiae* function as sensors include SNF3 and RGT2, which belong to the family of hexose transporters and can differentiate low and high concentrations of glucose in the extracellular medium (Conrad et al., 2014). The induction of genes which encode such precise carrier proteins, encourage efficient substrate import (Ozcan et al., 1999).

As hemicellulose is main substrate for xylitol production, hence it could be integrated with biorefinery for an economically feasible strategy in the large scale implementation of xylitol bioproduction (Yamakava et al., 2020). Additional valuable bioproducts can be produced via integral utilization of biomass which will promote faster growth of the biobased economy (Dragone et al., 2020). For this techno-economic studies have been done using various agro-industrial residues/by-products as raw materials such as sugarcane bagasse, corn stover, brewer's spent grains. Implementation of heat exchangers in an ethanol and xylitol biorefinery also brought about 68.9% energy saving as steam and 64.4% reduction in cooling water consumption Therefore, corn stover biorefinery for xylitol production integrated with second generation ethanol proved more economically feasible option (Franceschin et al., 2011).

Bioproduction of fatty alcohol

Higher order alcohols with amphipathic molecules i.e. presence of a carbon tail and non-ionizable hydroxyl head, are termed as fatty alcohols. The chain length vary from 6 to 26 carbons (C6–C26) (Kelm and Wickett, 2017). In 2017, its global market was around USD 6.9 billion and by 2023, it is estimated to be above USD 10 billion (Munkajohnpong et al., 2020). Fatty alcohols, have single bond and straight chain structure with amphiphilic nature, are effectively used in plasticizers, lubricants, detergents, additives in adhesives, ink and deink solutions, and personal care products (Borodina et al., 2018).

Production lies on chemical conversion of precursors from bio-based resources such as hydrogenation of oils, fats, methyl-and wax esters of fatty acids and through petro-based sources involving oligomerization of ethylene followed by the oxidation process (Yeong et al., 2012). Catalytic hydrogenation of fatty acid methyl ester (FAME), fatty acid wax ester (FAWE), and free fatty acid are involved in industries (Tabata et al., 1993). This process is classified as slurry-phase hydrogenation and fixed-bed hydrogenation (Hiles and Tuck, 1991; Sved et al., 1999). In slurry based hydrogenation direct suspension of catalysts occur in reactor while in fixed bed, gas or liquid reactants are passed through a column packed with immobilized catalysts.

Alternatively, bioprocessing involves engineering and incorporation of biosynthetic pathway into host cells to produce fatty alcohols, which reduce consumption of energy and heat, and chemical waste generation through the chemical process, apart from emissions of greenhouse gases (Garcia and You, 2015). It can be advantages in terms of environmental, social, and economic impacts (Gargalo et al., 2016).

Naturally, fatty alcohols are synthesized from fatty acids or fatty acid derivatives, fatty acyl-CoA or acyl carrier protein (ACP), through two major pathways, either a one- or two-enzyme reduction process. In single enzyme process, fatty acyl-CoA reductase (afFAR) catalyzes the reduction of fatty acyl-CoA/ACP using four electrons from 2 NADPH (Kolattukudy and Rogers, 1978) and forming fatty alcohol. While in two enzyme pathway, from fatty acyl-CoA, fatty acyl-ACP or fatty acid an intermediate called as fatty aldehyde intermediate is formed, which is further reduced to fatty alcohol by alcohol dehydrogenase (ADH) or aldehyde reductase (AHR). Two enzyme pathway is seen in natural microbes such as *Acinetobacter* sp. and cyanobacteria (Santala et al., 2014). Carboxylic acid reduction to aldehyde is achieved by enzyme carboxylic acid reductase (CAR). Other enzymes in aldehyde reductase family (AHR) are pyruvate formate-lyase deactivase (AdhE) and alcohol dehydrogenase (ADH) which is mainly ethanol dehydrogenase / alcohol dehydrogenase (AdhP) and converts fatty aldehyde to fatty alcohol (Schirmer et al., 2010).

Fatty alcohols are generated as intracellular products (Feng et al., 2015). Hence, recovery occurs through harvesting and lysing of host cells. This process involves filtration, flocculation and centrifugation, amongst which centrifugation is most common where without chemical contamination, microbial cells are collected (Wang et al., 2016). Due to large amount of energy consumption, filtration is also widely used. Cell disruption takes place by bead milling (Wang et al., 2016), temperature shock (Zhou et al., 2016) and ultrasonication (Mubarak et al., 2016). Followed by generally organic solvent (hexane, chloroform, ethyl acetate and methanol) treatment to recover the entire fraction of fatty alcohols from the whole cells in solvent. Since, in bioprocess, refining of fatty alcohol from metabolites, enzymes, salts, cell debris is required, which is achieved by evaporation / crystallization, fractional distillation, and solvent extraction.

Thus, bioprocessing is advantageous over chemical process as it consumes less energy, environment friendly, independent of petroleum resources, capable of using various substrates and can produce unsaturated alcohol also. The disadvantages include, slow process, low production yield, require sterilization and purification process. But in long term, it is going to be beneficial, and can achieve higher yield through genetic recombination techniques and optimization process (Munkajohnpong et al., 2020).

Butanol bioproduction

Butanol is mainly used in manufacture of plastics, paints and polymers, while recently considered as alternative fuel for transport in place of gasoline. It can be burned in gasoline engine, producing 25% more energy and lower gas emissions to the environment. Mainly it is made from crude oil through a petrochemical reaction (Li et al., 2019), however, its alternative biotechnological pathway is termed as ABE (acetone, butanol, ethanol). This is mainly achieved through fermentation by solventogenic strains of *Clostridium* sp. which generate ABE at a 3:6:1 ratio, respectively. In China, raw material for fermentation is corn and sugarcane, which attributes to 75% of final cost. Therefore, abundant lignocellulosic biomass is considered as alternative option these days (Volk and George, 2019). But the high cost of hydrolysing cellulose due to its recalcitrant behaviour is the main hurdle for industrial scale production. Lignocellulose is mainly composed of cellulose, which is homopolysaccharide of glucose monomers linked by β-1,4 glycosidic bonds, hemicellulose, which is heteropolysaccharide of multiple sugars mainly xylose and arabinose, and third component is lignin, which is polymer of phenolic units which are crosslinked among itself (Parisutham et al., 2017). Hence before production, the polymers have to be pretreated and enzymatically degraded to obtain sugars then the *C. acetobutylicum*, *C. beijerinckii*, *C. saccharoperbutylicum*, *C. pasteurianum*, *C. saccharoperbutylacetonicum* or *C. butylicum produce ABE from sugars* (Patakova et al., 2018).

The main glycolysis pathway produces acetyl-CoA from pyruvate, subsequently leading to acetoacetyl- CoA. Later, it forms -hydroxybutyryl-CoA which synthesizes crotonyl-CoA and butyryl-CoA. This leads to butyraldehyde formation which gets converted to butanol. In this pathway, acetyl-CoA and acetoacetyl- CoA gets converted to acetaldehyde and acetoacetate, which form ethanol and acetone, respectively (Ndaba et al., 2015). The enzymes necessary for starch degradation namely, α-amylases, α-glycosidases, glucoamylases, pullulanases and amylopullulanases are present in *Clostridium* sp. (Yang et al., 2017). Mainly the conventional process of butanol production is divided to three parts, i.e. physical, chemical, or biological pretreatment of biomass; followed by enzymatic hydrolysis (saccharification), and then fermentation ABE → acetone–butanol–ethanol (Gottumukkala et al., 2017). However, in consolidated bioprocessing (CBP) strategy, enzymatic hydrolysis and fermentation are carried out in one step (Ibrahim et al., 2018). The three enzyme systems for cellulose hydrolysis namely, endoglucanases, exoglucanases and β-glucosidases are located in multienzyme complexes known as cellulosomes in *C.cellulolyticum, C. cellulovorans, C. thermocellum, and C. celevecrescens* (Lin et al., 2015; Ou et al., 2017). In the cellulosome, enzymes are amassed in structural scaffolding subunits with the help of strong noncovalent interactions of proteins in the coupling (dockerin) and complementary (cohesins) modules. Hence a synergistic effect of enzyme substrate complex and spatial proximity of the different types of cellulases to each other result in efficient degradation of vegetative biomass (Xin et al., 2018). After fermentation, distillation is required to separate butanol from other solvents, hence it is the most technologically developed separation processes based on boiling point difference. It is the bottleneck for biobutanol production at an industrial scale, despite effective fermentation, since azeotropic mixture can be formed during the process. Therefore, more research is necessary to develop integrated processes that will allow combining one or more techniques to enhance butanol selectivity at lower costs (Jee and Lee, 2014).

Bioprocess control strategies

Control strategies of bioprocess involves a set of operations that has to supervise the unpredictable environment of the process with an aim to maintain the process within the desired design space. During the process development, control strategy is created. Deep understanding of the process is required to create robust control strategy and accurate process models. Inaccurate process models result in instability of the controller (Mears et al., 2017a). Major challenge is also of data availability as generally limited data is acquired during process development (Mears et al., 2017a; Yee et al., 2018). Based on the requirements, sophisticated instrumentation and complex control laws or simple control law having intricate monitoring system can be selected. Various control methods are as follows:

a. **Open Control Methods**- one of the oldest and simple technique used in fermentation process at industrial scale. Here generally online measurements not required and open loop strategy utilized to reduce batch to batch variability (Liu et al., 2019). Major drawbacks of this system are incompetency to take corrective action during random error while operation, requirement of complex database of the process, real time application challenge (Julia et al., 2019).

b. **Proportional Integral/ Proportional Integral Derivative (PI/PID) control**- They are closed loop control systems, where feedback term is added for regulatory action. PID controller has closed loop algorithm, which is known for ease in tuning, simplicity, and robustness. Recently, a SISO nonlinear bioreactor model using real-time data of *E.coli* fermentation had employed internal model control based PID (IMC-PID), which produced superior performance with minimum absolute error as compared to other methods (Joanofarc, 2019). Flow and/or pH feedback control loops have shown significant potential. PID control loops for microbial culture control also employed where, its composition is input variable and actuator influence population. Similarly, actuators can be used for temperature, dissolved oxygen and pH, as it gives high reproducibility of processes with quality product (Simutis and Lübbert, 2015). Its limitation is its inability to cope up with non-linear processes (Pantano et al., 2017). As the underlying physicochemical processes are dynamic and complex, hence controller needs to be modified for non-linear processes also.

c. **Cascade Control-** For improved dynamic responses of the system, cascade controls are widely used in industry. Effect of load disturbance is minimised in the secondary loop. It involves multiple sensors for measuring conditions in controlled process, hence better than conventional single loop control (Campos-Rodríguez et al., 2019). It is one step superior to PI/PID control as it allows satisfactory control output and secondary variable measurement (Raja and Ali, 2017). It nullifies batch variation by controlling short-time process disturbances (Kuprijanov et al., 2011). Advantages of this system include, distribution of disturbance in secondary loop where corrective measures are taken, lag phase is totally abated in secondary loop, thus improving response, secondary loop accurately regulated mass/energy flow. Hence this control system can be effectively used in biotherapeutics production.

d. **Model Predictive Control (MPC)-** It is comprehensively used for bioprocess optimization. A predictive process model governs the dynamic and static interactions between input, output and disturbance variables generating optimum set points calculations (Bolton, 2021). Non-linear MPC (NMPC) with larger computational time successfully used in complex mammalian bioprocesses. Its success mainly depends on model accuracy and ability to handle system disturbances (Mears et al., 2017b). MPC is computationally expensive when optimization is required at each step, therefore, its acceptance in biotech field require development of robust process models.

e. **Neural Network-Based Control-** NN has adaptive learning and effectively deals with non-linear systems. Data availability is biggest limitation, which is overcome by generating artificial datasets (random noise to original datasets. For tracking control, adaptive neural network is used (Natarajan et al., 2021). A study done on predicting fungal biomass by Multiphase Artificial Neural Network (MANN) is found to be successful as monitoring transient state is also possible (López et al., 2017). They are more descriptive than empirical models, their structure need not to be predefined, have more flexibility in modelling and are less noise prone.

f. **Fuzzy Logic-Based Control**- This logic integrates human experience and reasoning for designing nonlinear controllers. They have more practical applications, especially in fermentation bioprocess laden with several uncertainty factors which require optimization of process parameters for higher growth and productivity (Abyad et al., 2017). It has low mean square errors in nonlinear systems. Fuzzy feedforward control mechanism for temperature control or product concentration in fermentation or enzyme reactor respectively, has been proved to be efficient in non-linear process. Combination of fuzzy logic for controlling substrate concentration and ANN as soft sensor for glucose concentration estimation had more efficiency and less error (6%) (Escalante-Sánchez et al., 2018). The limitation of fuzzy system is inaccurate parameter estimation because of fine tuning of rules, lack of adaptability for dynamic process or minute change in variables may have great impact on process. Continuous and independent optimization of operating parameters is difficult.

g. **Model-Based Control**- Process models can be empirical, mathematical or statistical for monitoring, optimization and control (Benattia et al., 2020). DO-stat or pH stat control strategies are based on indirect feedback control having simple on-off controller. This control method used in manufacturing monoclonal antibodies, where upstream and downstream process included protein A capture, viral activation, ion exchange chromatography integrated together (Gomis-Fons, 2020). Realization of process control and supervision is based on mathematical model.

h. **Soft Sensor Based Control**- Generally process parameters are evaluated at-line or off-line leading to system cost and inefficient information delivery, hence integrated physical system are controlled by soft sensors and algorithms in an on-line mode. Their development is vast and requires non-invasive analytical techniques for real-time implementation (Steinwandter et al., 2019).

i. **Automated Biomanufacturing**- Digitization with online smart sensors, wireless technologies, advancements in sensor calibration and compact technologies are being increasingly used. For integrating unit operations, object linking and embedding approach is implemented. Automated and isolated workstations comprise consistency between software versions, proprietary interfaces and data formats, using standardized communication protocols and graphical user interphases (Chopda et al., 2016). The challenges for implementing plant-wide control panel are shop floor control systems (SFC), manufacturing execution systems (MES), integration of enterprise control system (ECS) into business systems (BS). Centralized distributed hardware/ software automation structure such as supervisory control and data acquisition (SCADA) platform has end-to end integrated system for analysing various unit operations (Feidl et al., 2020).

Data handling is also big challenge as its quantity and complexity requires operable industry standard software with high agility and flexibility. Biomanufacturing automation acceptance has only one crucial technical barrier i.e. bioprocess analytics (Charan and Gupta, 2016).

Bioprocess Microfluidics

An in-depth understanding of the bioprocesses is the basic requirement for enabling their rapid and successful development and implementation. Relevant data acquisition with minimum amount of labour and at minimum cost. Usually, over 10 000 experiments are necessary for a single bioprocess development project, from primary strain screening to pilot-scale trials in which key experiments are performed with bench-scale bioreactors (Wewetzer et al., 2015). Hence, it is an expensive and labor intensive endeavor. Therefore, miniaturized bioreactors, with volumes typically below 100 mL, having both stirred tank reactors and shaken systems are alternatives to bench-scale bioreactors. Automation of the optimization processes is enabled through microtiter plates with simplicity and operate with very small volumes. In combination with design of experiments (DoE) approaches, the total number of experiments required is reduced and can be less expensive (Marques and Szita, 2017). Microfluidic devices in which small amounts of fluids (10^{-9} and 10^{-18} l) are manipulated have emerged for high throughput applications (Marques and Szita, 2016). They can also be operated as batch, fed-batch, and continuous modes under a broad range of process conditions. There is laminar flow which gives control over temporal and spatial microenvironment beneficial for controlled delivery of reagents, cells or enzymes. Key process variables in these devices are monitored through optical sensors and luminescence quenching principles (oxygen, pH, CO2, glucose and temperature) by spectroscopy methods (e.g. NIR and Raman) (Kirk and Szita., 2013; Schäpper et al., 2009) and by image analysis (Jaccard et al., 2014). These devices were earlier used for the study of cellular microenvironments of stem cells in a reproducible and controllable fashion. Nowadays, with the incorporation of analytical tools in microfluidic cell culture devices (μCCD), they are used in several bioprocesses also. These tools need to be non-invasive to allow on-line monitoring to capture the dynamic behavior of the cells. These are generally light microscopy (Sbalzarini, 2016, optical sensors (Gruber et al., 2017) and electric cell-substrate impedance sensing (ECIS) (Xu et al., 2016b). For further knowledge, these on-line methods are complemented with off-line or at-line methods, such as FACS, HPLC and flow cytometry.

Kostov et al. worked in microbioreactors (μBR), with the integration of optical sensors to monitor *Escherichia coli* fermentations which have working volumes in the submillitre range (Kostov et al., 2001). Advances in polymer microfabrication and sensor miniaturization, have developed a large number of μBRs since then (Gruber et al., 2017). Exquisite gas permeability of poly-dimethylsiloxane (PDMS) is useful for the aeration of the culture, and structuring polymers like poly (methyl methacrylate) (PMMA) or PDMS to create fluidic structures, works well with μBR (Tsao., 2016). μBRs have optical sensors for monitoring as they have small footprint and are non-destructive. For industrial uptake, additional developments in sensing technology are required to detect nutrient consumption and product formation (Gruber et al., 2017). This will permit mathematical model integration and Quality by Design (QbD) approaches for bio-process development. Online monitoring of nutrients and product can be achieved with fluorescence-based sensors, surface plasmon resonance, Raman spectroscopic methods. Device components need to be standardized for automated workflows and process integration in particular for chromatography or mass spectrometry (Gernaey et al., 2012). These devices can be helpful in downstream processing (μDSP) which can deliver scale-relevant or scale-translatable data. But there is difficulty to miniaturize traditional DSP system (owing to complex geometries of the industrial scale equipment) (Titchener-Hooker, 2008). μDSP is recently used in process analysis of solid–liquid based separations (Tetala and Vijayalakshmi, 2016; Huft et al., 2013) by adsorptive techniques or use of membranes, liquid–liquid separations (Heintz et al., 2017) and crystallization (Shi et al., 2017). μBR application largely depends on the integration of analytics and the amount of samples accessible for quality control purposes. μDSP integration with upstream devices will create significant impact in minimizing development time and expenses of full-scale processes and eventually have increased value when complex therapies require processing (Rayat et al., 2016).

Bioprocessing with supercritical fluids

Unicellular, photosynthetic microorganisms living in marine or freshwater environments are referred to as microalgae. They transform carbon dioxide, water and light into new algal biomass. They have complex chemical composition. The concept of algae-based biorefinery refers to the comprehensive process optimization from biomass production to production of wide range of biochemical products. This requires first growing and harvesting of microalgae biomass, later followed by extraction and downstream processing. The biomass is composed of pigments, carotenoids, chlorophylls, polyphenol, sterols, fatty acids, lipids, proteins, polysaccharides, vitamins and minerals. These have abundant applications in the food, feed, pharmaceutical, cosmetic and bioenergy industries (Michalak et al., 2015). Hence, sustainable and feasible industrial scale production is the requirement. The first operations in downstream processing of microalgae biomass, pre-treatment, such as drying and milling have great impact on efficiency due to thick and hard cell wall of microalgae having hydrocarbon composition, that can hinder the extraction of internal metabolites. Therefore, supercritical fluid extraction (SFE) successfully used on *Botryococcus braunni, Chlorella vulgaris, Dunaliella salina, Haematococcus pluvialis, Nannochloropsis* sp., *Neochloris, Scenedesmus* and *Spirulina* sp. Advantages related to supercritical fluids (SCFs) are its transport properties, high diffusivity and low viscosity facilitate mass transfer phenomena, variation of pressure and/or temperature are adjusted providing high flexibility in the extraction. Carbon dioxide (SC-CO2) is the most commonly used SCF at laboratorial and industrial scale, because of low cost, non-toxicity, non-flammability, inertness, possibility of total recovery and moderate critical properties (P_c = 7.38 MPa, T_c = 304.2 K) compared to other green solvents (Reverchon and Marco, 2006; Coelho and Palavara, 2015). SFE of lipophilic compounds from microalgae have been carried out using CO2, as well as CO2 modified with polar co-solvents (e.g. ethanol), which have achieved very high yields of extraction, alongwith substantially improved selectivity and good fractionation capability (Ruen-Nagam et al., 2012; Nobre et al., 2013).

Microalga *Scenedesmus* sp. has been utilized for extraction of neutral lipids and fatty acids at 150 bar and 25°C and with the addition of methanol as entrainer. The pre-treatment of algal biomass executed using microwave radiation, gridding with liquid N2, ultrasonication, freeze-drying and osmotic shock and cooling. It was confirmed that microwave radiation (in the presence of water) exhibited highest potential to release lipids and fatty acids from the microalga biomass, with an extraction yield of 9.6% (wt.) could be reached (Viner et al., 2018). Extraction of lutein from *Chlorella pyrenoidosa* involving ultrasound-enhanced subcritical CO2 extraction by researchers (Fan et al., 2018). Extraction was enhanced using lower temperature and pressure (27°C and 21 MPa) compared to other methods. For minimizing the drawbacks associated to conventional extraction processes pressurized fluids have been employed on plant matrices to obtain high value-added compounds. These are mainly pressurized fluid extraction (PFE), pressurized liquid extraction (PLE), pressurized hot solvent extraction (PHSE), accelerated solvent extraction (ASE), high-pressure, high temperature solvent extraction (HPHTSE), high-pressure solvent extraction (HPSE) and subcritical solvent extraction (SSE) (Benthin et al., 1999; Mustafa and Turner, 2011; Herrero et al., 2005). Involvement of both high temperature and pressure of the solvent brings higher solubility of the solutes, at low viscosity as well as surface tension with an improved penetration in the matrix and leading to higher mass transfer rates. Solvent is always below the critical point so that the liquid state is maintained. The equipment operates at static mode and automated. The parameters which influence are extraction temperature, number of cycles, time of static cycle, particle size and water content of the sample (Mustafa and Turner, 2011; Alonso-Salces et al., 2001; Ramos et al., 2002). Only GRAS solvents are employed in PLE such as water and ethanol. Thus, SFE has lower environmental impact and economically feasible.

References

Abyad, M., Karama, A. & Khallouq, A. (2017). Fuzzy Takagi-Sugeno based modelling and control for an alcoholic fermentation process. In Proceedings of the 2017 International Conference on Electrical and Information Technologies (ICEIT), Rabat, Morocco, 15–18, November 2017; Institute of Electrical and Electronics Engineers (IEEE): Piscataway, NJ, USA, 1–6.

Alonso-Salces R.M., Korta E., Barranco A., Berrueta L.A., Gallo B., Vicente F. (2001). Pressurized liquid extraction for the determination of polyphenols in apple. J Chromatogr A, 933:37–43.

Arcaño Y.D., Valmaña García O.D., Mandelli D., Carvalho W.A., Magalhães Pontes L.A. (2020). Xylitol: a review on the progress and challenges of its production by chemical route. Catal Today, 344:2–14.

Barik S. & Paul K.K. (2017). Potential reuse of kitchen food waste. Journal of Environmental Chemical Engineering, 5,196-204.

Bernstad, A. & Jansen J. L. C. (2012). Review of comparative LCAs of food waste management systems –Current status and potential improvements. Waste Management. 32, 2439-55.

Basu S. & Yang S.T. (2005). Bioseparations, in S. Lee (ed.), Encyclopedia of Chemical Processing,Marcel Dekker, 221-236.

Benattia, S.E., Tebbani, S. & Dumur, D. (2020). Linearized min-max robust model predictive control: Application to the control of a bioprocess. Int. J. Robust Nonlinear Control, 30, 100–120.

Benthin B., Danz H. & Hamburger M. (1999). Pressurized liquid extraction of medicinal plants. J Chromatogr A, 837:211–219.

Bolton, W. (2021). Controllers. In Instrumentation and Control Systems, 3rd ed.; Elsevier Science: Amsterdam, The Netherlands, 297–327, Chapter 13.

Borodina I., Holkenbrink C., Dam M.I., & Loefstedt C. (2018). Inventors; Danmarks Tekniske Universitet, assignee. Methods for producing fatty alcohols and derivatives thereof in yeast. Denmark patent WO2018109163A1. 15 December 2017

Camire J, et al. Adherent Growth of CHO cells in ThermoScientific cyClone SFM4CHO-A medium using Corning® CellBIND® Surface Culture Flasks. Application Note 1.0. 2007.

Campos-Rodríguez, A., García-Sandoval, J. &González-Álvarez, V. (2019). Hybrid cascade control for a class of nonlinear dynamical systems. J. Process. Control, 76, 141–154.

Carvalheiro F., Duarte L.C., Gírio F., Moniz P. (2016). Hydrothermal/liquid hot water pretreatment (autohydrolysis): a multipurpose process for biomass upgrading. https://doi.org/10.1016/B978-0-12-802323-5.00014-1.

Charan, H.Y. & Gupta, N.V. (2016). Gamp 5 a Quality Risk Management Approach to Computer System Validation. Int. J. Pharm. Sci. Rev. Res, 36, 195–198.

Chatzifragkou A., Papanikolaou S., Kopsahelis N., Kachrimanidou V., Dorado M.P., & Koutinas A.A. (2014). Biorefinery development through utilization of biodiesel industry by-products as sole fermentation feedstock for 1,3-propanediol production. Bioresource Technology. 159, 167-75.

Cho D.H., Lee Y.J., Um Y., Sang B.I., Kim Y.H. (2009). Detoxification of model phenolic compounds in lignocellulosic hydrolysates with peroxidase for butanol production from Clostridium beijerinckii. Appl Microbiol Biotechnology, 83: 1035–43.

Chopda, V.R., Gomes, J. & Rathore, A.S. (2016). Bridging the gap between PAT concepts and implementation: An integrated software platform for fermentation. Biotechnol. J. 11, 164–171.

Coelho J.P., Palavra A.F. (2015). Supercritical fluid extraction of compounds from spices and herbs. In: Fornari T, Stateva RP, editors. High pressure fluid technology for green food processing. New York Dordrecht London: Cham Heidelberg, 357–96.

Conrad M., Schothorst J., Kankipati H.N., Van Zeebroeck G., Rubio-Texeira M., Thevelein J.M. (2014). Nutrient sensing and signaling in the yeast *Saccharomyces cerevisiae*. FEMS Microbiol Rev,38:254–99.

Dorado M.P., Lin S.K.C., Koutinas A., Du C., Wang R. & Webb C. (2009). Cereal-based biorefinery development:Utilisation of wheat milling by-products for the production of succinic acid. Journal of Biotechnology. 143, 51-9.

Dorceus M. Cell Culture Scale-Up in Stirred-Tank Single-Use Bioreactors. BioProcess International. 16(11–12) si. 2018.

Dragone G., Kerssemakers A.A.J., Driessen J.L.S.P., Yamakawa C.K., Brumano L.P., Mussatto S.I. (2020). Innovation and strategic orientations for the development of advanced biorefineries. Bioresour Technol, 302:122847.

Elbeshbishy E., Dhar B.R., Nakhla G. & Lee H.S. (2017). A critical review on inhibition of dark biohydrogen fermentation. Renewable & Sustainable Energy Reviews, 79.656–68.

Elbing K., Larsson C., Bill R.M., Albers E., Snoep J.L., Boles E., et al. (2004). Role of hexose transport in control of glycolytic flux in Saccharomyces cerevisiae. Appl Environ Microbiol, 70:5323–30.

Escalante-Sánchez, A., Barrera-Cortés, J., Poggi-Varaldo, H.M., Ponce-Noyola, T. & Baruch, I.S. (2018). A soft sensor based on online biomass measurements for the glucose estimation and control of fed-batch cultures of *Bacillus thuringiensis*. Bioprocess Biosyst. Eng, 41, 1471–148.

Ewert J., Glück C., Strasdeit H., Fischer L. & Stressler T. (2018). Influence of the metal ion on the enzyme activity and kinetics of PepA from *Lactobacillus delbrueckii*. Enzyme and Microbial Technology, 110,69-78

Fan X.D., Hou Y., Huang X.X., Qiu T.Q. & Jiang .JG. (2015). Ultrasound-enhanced subcritical CO2 extraction of lutein from *Chlorella pyrenoidosa*. J Agric Food Chem, 63:4597–05.

Fancer K.A., Rohani R., Mohammad A.W., Ba-Abbad M.M. (2017). Evaluation of the operating parameters for the separation of xylitol from a mixed sugar solution by using a polyethersulfone nanofiltration membrane. Kor J Chem Eng, 34: 2944–57.

Feidl F., Vogg S., Wolf M., Podobnik M., Ruggeri C., Ulmer N., Wälchli R., Souquet J., Broly H., Butté A., et al. (2020). Process-wide control and automation of an integrated continuous manufacturing platform for antibodies. Biotechnol. Bioeng, 117, 1367–1380.

Feng X., Lian J. & Zhao H. (2015). Metabolic engineering of *Saccharomyces cerevisiae* to improve 1-hexadecanol production. Metab Eng 27:10–19.

Franceschin G., Sudiro M., Ingram T., Smirnova I., Brunner G., Bertucco A. (2011). Conversion of rye straw into fuel and xylitol: a technical and economical assessment based on experimental data. Chem Eng Res Des, 89:631–40.

Galanakis C.M. (2015). Chapter 3 - The universal recovery strategy. Food Waste Recovery. Academic Press, San Diego, 59-81.

Gallezot P. (2013). Metal catalysts for the conversion of biomass to chemicals. Elsevier B. V.

Gänzle M.G., Vermeulen N. & Vogel R.F. (2007). Carbohydrate, peptide and lipid metabolism of lactic acid bacteria in sourdough. Food Microbiology, 24,128-38.

Garcia D.J. & You F. (2015). Multiobjective optimization of product and process networks: general modeling framework, efficient global optimization algorithm, and case studies on bioconversion. AIChE J 61(2):530–554.

Gargalo C.L., Cheali P., Posada J.A., Carvalho A., Gernaey K.V. & Sin G. (2016). Assessing the environmental sustainability of early stage design for bioprocesses under uncertainties: an analysis of glycerol bioconversion. J Clean Prod 139:1245– 1260.

Gernaey K.V., Baganz F., Franco-Lara E., Kensy F., Kruehne U., Luebberstedt M., Marx U., Palmqvist E., Schmid A., Schubert F. et al. (2012). Monitoring and control of microbioreactors: an expert opinion on development needs. Biotechnol J, 7:1308-1314.

Gomis-Fons, J., Schwarz, H., Zhang, L., Andersson, N., Nilsson, B., Castan, A., Solbrand, A., Stevenson, J., Chotteau, V. (2020). Model based design and control of a small-scale integrated continuous end-to-end mAb platform. Biotechnol. Prog, 36, e2995.

Gottumukkala L.D., Haigh K. & Görgens, J. (2017). Trends and advances in conversion of lignocellulosic biomass to biobutanol: microbes, bioprocesses and industrial viability. Renew. Sustain. Energy Rev, 76, 963–973.

Gruber P., Marques M.P.C., Szita N., Mayr T. (2017). Integration and application of optical chemical sensors in microbioreactors. Lab Chip, 17:2693-2712.

Heintz S., Börner T., Ringborg R.H., Rehn G., Grey C., Nordblad M., Krühne U., Gernaey K.V., Adlercreutz P., Woodley J.M. (2017). Development of in situ product removal strategies in biocatalysis applying scaled-down unit operations. Biotechnol Bioeng, 114: 600-609.

Hernández-Pérez A.F., Arruda P.V., Sene L., Silva S.S., Chandel A.K., Felipe M.G.A. (2019). Xylitol bioproduction: state-of-the-art, industrial paradigm shift, and opportunities for integrated biorefineries. Crit Rev Biotechnology, 39:924–43.

Herrero M., Martín-Álvarez P.J., Señoráns F.J., Cifuentes A. Ibáñez E. (2005). Optimization of accelerated solvent extraction of antioxidants from *Spirulina platensis* microalga. Food Chem, 93:417–23.

Hess G. (2006). Push for biofuels seen in farm bill, Chem Eng News, May 22, 29-31

Hiles A.G. & Tuck M.W. (1991). inventors; Davy McKee (London) Limited, Assignee. Hydrogenation Process. USA, July 27, 1990.

Huft J., Haynes C.A. & Hansen C.L. (2013). Microfluidic integration of parallel solid-phase liquid chromatography. Anal Chem, 85:2999-3005.

Ibrahim M.F., Ramli N., Bahrin E.K. & Abd-Aziz S. (2017). Cellulosic biobutanol by *Clostridia*: challenges and improvements. Renew. Sust. Energy Rev, 79, 1241– 1254.

Jaccard N., Griffin L.D., Keser A., Macown R.J.., Super A, Veraitch F.S., Szita N. (2014). Automated method for the rapid and precise estimation of adherent cell culture characteristics from phase contrast microscopy images. Biotechnol Bioeng, 111: 504-517.

Jee K.Y. & Lee Y.T. (2014). Preparation and characterization of siloxane composite membranes for n-butanol concentration from ABE solution by pervaporation. J. Membr. Sci., 456, 1–10.

Joanofarc, X., Nivedhika, D., Patnaik, S. & Panda, R. (2019). Closed-loop Performance and Analysis of a Real Time Non-linear Bioreactor Process. In Proceedings of the 2019 2nd International Conference on Power and Embedded Drive Control (ICPEDC), Chennai, India, 21–23 August 2019; Institute of Electrical and Electronics Engineers (IEEE): Piscataway, NJ, USA, 406–412.

Julia, R., Xavier, G., José, L.M., Denise, F. & Francisco, V. (2019). Continuous operation, a realistic alternative to fed-batch fermentation for the production of recombinant lipase B from *Candida antarctica* under the constitutive promoter PGK in *Pichia pastoris*. Biochem. Eng. J, 147, 39–47.

Kelm G.R. & Wickett R.R. (2017).The role of fatty acids in cosmetic technology, in Fatty Acids, ed. by Ahmad MU. AOCS Press, Washington, DC, 385–404.

Kim S., Lee J., & Sung B.H. (2019). Isolation and characterization of the stress-tolerant *Candida tropicalis* YHJ1 and evaluation of its xylose reductase for xylitol production from acid pre-treatment wastewater. Front Bioeng Biotechnol,7: 138.

Kirk T.V., & Szita N. (2013). Oxygen transfer characteristics of miniaturized bioreactor systems. Biotechnol Bioeng, 110:1005-1019.

Ko B.S., Kim J. & Kim J.H. (2006). Production of xylitol from D-xylose by a xylitol dehydrogenase gene-disrupted mutant of. Candida tropicalis. Appl Environ Microbiol, 72:4207–13.

Kolattukudy P.E. & Rogers L. (1978). Biosynthesis of fatty alcohols, alkane-1,2-diols and wax esters in particulate preparations from the uropygial glands of white-crowned sparrows (*Zonotrichia leucophrys*). Arch Biochem Biophys 191(1): 244–258.

Koppram R., Albers E. & Olsson L. (2012). Evolutionary engineering strategies to enhance tolerance of xylose utilizing recombinant yeast to inhibitors derived from spruce biomass. Biotechnol Biofuels, 5:1–12.

Kostov Y., Harms P., Randers-Eichhorn L. & Rao G. (2001). Low-cost microbioreactor for high-throughput bioprocessing. Biotechnol Bioeng, 72:346-352.

Kumar V., Sandhu P.P., Ahluwalia V., Mishra B.B. & Yadav S.K. (2019). Improved upstream processing for detoxification and recovery of xylitol produced from corncob. Bioresour Technol, 291:121931.

Kumar J.V., Mathew R. & Shahbazi A. (1998). Bioconversion of solid food wastes to ethanol. Analyst. 123, 497-502.

Kuprijanov, A., Schaepe, S., Aehle, M., Simutis, R. & Lübbert, A. (2011). Improving cultivation processes for recombinant protein production. Bioprocess. Biosyst. Eng, 35, 333–340.

Li Y., Tang W., Chen Y., Liu J. & Lee C.F. (2019). Potential of acetone-butanol-ethanol (ABE) as a biofuel. Fuel, 242, 673–686.

Lin P.P., Mi L., Morioka A.H., Yoshino K.M., Konishi S., Xu S.C., Papanek B.A., Riley L.A., Guss A.M. & Liao, J.C. (2015). Consolidated bioprocessing of cellulose to isobutanol using *Clostridium thermocellum*. Metab. Eng, 31, 44–52.

Liu, W.-C., Inwood, S., Gong, T., Sharma, A., Yu, L.-Y. & Zhu, P. (2019). Fed-batch high-cell-density fermentation strategies for *Pichia pastoris* growth and production. Crit. Rev. Biotechnol, 39, 258–271.

López, M.E., Rene, E.R., Boger, Z., Veiga, M.C. & Kennes, C. (2017). Modelling the removal of volatile pollutants under transient conditions in a two-stage bioreactor using artificial neural networks. J. Hazard. Mater., 324, 100–109.

Maina S., Pateraki C., Kopsahelis N., Paramithiotis S., Drosinos E.H., Papanikolaou S., et al. (2017). Microbial oil production from various carbon sources by newly isolated oleaginous yeasts. Engineering in Life Sciences. 17, 333-44.

Marques M.P.C. & Szita, N. (2017). Bioprocess microfluidics: applying microfluidic devices for bioprocessing. Current Opinion in Chemical Engineering, 18:61–68

Marques M.P.C, Szita N. (2016). Microfluidic devices for the culture of stem cells. In Stem Cell Manufacturing. Edited by Cabral JM, Silva CLd., Chase LG, Diogo MMBT. Elsevier;:171-198.

Marton J.M., Felipe M.G.A., Almeida e Silva J.B., Pessoa A. (2006). Evaluation of the activated charcoals and adsorption conditions used in the treatment of sugarcane bagasse hydrolysate for xylitol production. Braz J Chem Eng; 23:9–21.

Mears, L., Stocks, S.M., Albaek, M.O., Sin, G. & Gernaey, K.V. (2017a). Mechanistic fermentation models for process design, monitoring, and control. Trends Biotechnol., 35, 914–924.

Mears, L., Stocks, S.M., Sin, G. & Gernaey, K.V. (2017b). A review of control strategies for manipulating the feed rate in fed-batch fermentation processes. J. Biotechnol, 245, 34–46.

Michalak I., Dmytryk A., Wieczorek P.P., Rój E., Łęska B., Górka B., et al. (2015). Supercritical algal extracts: a source of biologically active compounds from nature. J Chem, 2015:1–14.

Misra S., Gupta P., Raghuwanshi S., Dutt K., Saxena R.K. (2011). Comparative study on different strategies involved for xylitol purification from culture media fermented by *Candida tropicalis*. Separ Purif Technol, 78:266–73.

Mubarak M., Shaija A. & Suchithra T.V. (2016). Ultrasonication: an effective pre-treatment method for extracting lipid from *Salvinia molesta* for biodiesel production. Resour Efficient Technol 2(3):126–132.

Munkajohnpong P., Kesornpun C., Buttranon S., Jaroensuk J., Weeranoppanant N., & Chaiyen P. (2020). Fatty alcohol production: an opportunity of bioprocess: Review. Biofuels, Bioprod. Bioref. DOI: 10.1002/bbb.2112

Mussatto S.I. (2016). Biomass fractionation technologies for a lignocellulosic feedstock based biorefinery. Elsevier.

Mussatto S.I. (2012). Application of xylitol in food formulations and benefits for health. In: Silva SS, Chandel AK, editors. D-xylitol ferment. Prod. Appl. Commer. Berlin, Heidelberg: Springer Verlag, 309–23.

Mustafa A. & Turner C. (2011). Pressurized liquid extraction as a green approach in food and herbal plants extraction: A review. Anal Chim Acta, 703:8–18.

Natarajan, P., Moghadam, R. & Jagannathan, S. (2021). Online deep neural network-based feedback control of a Lutein bioprocess. J. Process. Control, 98, 41–51.

Ndaba B., Chiyanzu I. & Marx S. (2015). N-Butanol derived from biochemical and chemical routes: a review. Biotechnol. Rep, 8, 1–9.

Nelson D.L., Nelson D.L., Lehninger A.L., Cox. M.M. (2008). Lehninger principles of biochemistry. W.H. Freeman, New York.

Nguyen N., Fargues C., Guiga W., Lameloise M.L. (2015). Assessing nanofiltration and reverse osmosis for the detoxification of lignocellulosic hydrolysates. J Membr Sci, 487:40–50.

Nobre B.P., Villalobos F., Barragán B.E., Oliveira A.C., Batista A.P., Marques P.A.S., et al. (2013). A biorefinery from *Nannochloropsis* sp. microalga - Extraction of oils and pigments. Production of biohydrogen from the leftover biomass. Bioresour Technol.135:128–36.

Oak Ridge National Laboratory. (2005). Biomass as Feedstock for a Bioenergy and BioproductsIndustry: The Technical Feasibility for a Billion-Ton Annual Supply, U.S. Department of Energy, DOE/GO-201995-2135.

Ou J., Xu N., Ernst P., Ma C., Bush M., Goh K.Y., Zhao J., Zhou L., Yang S.T. & Liu X. (2017). Process engineering of cellulosic n-butanol production from corn-based biomass using *Clostridium cellulovorans*. Process Biochem, 62, 144–150.

Ozcan S. & Johnston M. (1999). Function and regulation of yeast hexose transporters. Microbiol Mol Biol Rev, 63:554–69.

Pantano, M.N., Serrano, M.E., Fernández, M.C., Rossomando, F.G., Ortiz, O.A. & Scaglia, G.J.E. (2017). Multivariable control for tracking optimal profiles in a nonlinear fed-batch bioprocess integrated with state estimation. Ind. Eng. Chem. Res., 56, 6043–6056.

Papanikolaou S. & Aggelis G. (2011). Lipids of oleaginous yeasts. Part I: Biochemistry of single cell oil production. Europen Journal of Lipids Science Technology, 113, 1031-51.

Parisutham V., Chandran S.P., Mukhopadhyay A., Lee S.K. & Keasling J.D. (2017). Intracellular cellobiose metabolism and its applications in lignocellulose-based biorefineries. Bioresour. Technol, 239, 496–506.

Park J.B., Kim J.S., Kweon D.H., Kweon D.H., Seo J.H., Ha S.J. (2019). Overexpression of endogenous xylose reductase enhanced xylitol productivity at 40 ◦C by thermotolerant yeast *Kluyveromyces marxianus*. Appl Biochem Biotechnol, 189:459–70.

Patakova P., Kolek J., Sedlar K., Koscova P., Branska B., Kupkova K., Paulova L. & Provaznik, I. (2018). Comparative analysis of high butanol tolerance and production in *Clostridia*. Biotechnol. Adv, 36, 721–73

Perez, J., Munoz-Dorado, J., de la Rubia, T., & Martinez, J. (2002). Biodegradation and biological treatments of cellulose, hemicellulose and lignin: an overview, Int Microbiol 5, 53–63.

Petri R. and Schmidt-Dannert C. (2004). Dealing with complexity: evolutionary engineering and genome shuffling. Curr. Opin. Biotechnol. 15, 298-304.

Procentese A., Raganati F., Olivieri G., Russo M.E., Rehmann L. & Marzocchella A. (2018). Deep eutectic solvents pretreatments of agro-industrial food waste. Biotechnology & Biofuels, 11, 37.

Queiroz S.S., Jofre F.M., Mussatto S.I., Felipe M.D.G.A. (2022). Scaling up xylitol bioproduction: Challenges to achieve a profitable bioprocess. Renewable and Sustainable Energy Reviews 154, 111789

Raja, G.L. & Ali, A. (2017). Series cascade control: An outline survey. In Proceedings of the 2017 Indian Control Conference (ICC), Guwahati, India, 4–6 January 2017; Institute of Electrical and Electronics Engineers (IEEE): Piscataway, NJ, USA, 409–414.

Ramos L., Kristenson E.M., Brinkman U.A. (2002). Current use of pressurised liquid extraction and subcritical water extraction in environmental analysis. J Chromatogr A, 975:3–29.

Rao L.V., Goli J.K., Gentela J., Koti S. (2016). Bioconversion of lignocellulosic biomass to xylitol: an overview. Bioresour Technol,213:299–310.

Rayat A.C., Chatel A., Hoare M., Lye G.J. (2016). Ultra scale-down approaches to enhance the creation of bioprocesses at scale: impacts of process shear stress and early recovery stages. Curr Opin Chem Eng, 14:150-157.

Research & Markets Reports. Xylitol - global market trajectory & analytics. 2020. https://www.researchandmarkets.com/reports/5140553/xylitol-global-market-trajectory-and-analytics. [Accessed 22 April 2021].

Reverchon E. & De Marco I. (2006). Supercritical fluid extraction and fractionation of natural matter. J Supercrit Fluids.38:146–66.

Ruen-Ngam D., Shotipruk A., Pavasant P., Machmudah S., Goto M. (2012). Selective extraction of lutein from alcohol treated *Chlorella vulgaris* by supercritical CO2. Chem Eng Technol.35:255–60.

Ruijter G.J.G., Visser J. & Rinzema A. (2004). Polyol accumulation by *Aspergillus oryzae* at low water activity in solid-state fermentation. Microbiology. 150, 1095-101.

Sajjadi B., Raman A.A.A. &Arandiyan H. (2016). A comprehensive review on properties of edible and non-edible vegetable oil-based biodiesel: composition, specifications and prediction models. Renewable & Sustainable Energy Reviews, 63, 62-92.

Santala S., Efimova E., Koskinen P., Karp M.T. & Santala V. (2014). Rewiring the wax ester production pathway of *Acinetobacter baylyi* ADP1. ACS Synth Biol 3(3):145–151.

Santos J.C., Marton J.M., Felipe M.G.A. (2014). Continuous system of combined columns of ion exchange resins and activated charcoal as a new approach for the removal of toxics from sugar cane bagasse hemicellulosic hydrolysate. Ind Eng Chem Res, 53:16494–501.

Sbalzarini I.F. (2016). Seeing is believing: quantifying is convincing: computational image analysis in biology. In Advances in Anatomy, Embryology, and Cell Biology. Edited by De Vos W, Munck S, Timmermans J. Cham: Springer:1-39.

Schäpper D., Alam M.N.H.Z., Szita N., Eliasson Lantz A., Gernaey K.V., Schapper D., Alam M.N.H.Z., Szita N., Lantz A.E., Gernaey K.V. (2009). Application of microbioreactors in fermentation process development: a review. Anal Bioanal Chem, 395:679-695.

Sharma N.K., Behera S., Arora R., Kumar S. (2016). Enhancement in xylose utilization using *Kluyveromyces marxianus* NIRE-K1 through evolutionary adaptation approach. Bioproc Biosyst Eng, 39:835–43.

Shi H., Xiao Y., Ferguson S, Huang X., Wang N., Hao H. (2017). Progress of crystallization in microfluidic devices. Lab Chip http://dx. doi.org/10.1039/C6LC01225F.

Shirai, K., Guerrero I., Huerta S., Saucedo G., Castillo A., Obdulia Gonzalez R., et al. (2001). Effect of initial glucose concentration and inoculation level of lactic acid bacteria in shrimp waste ensilation. Enzyme and Microbial Technology. 28, 446-52.

Schirmer A., Rude M.A., Li X., Popova E. & del Cardayre S.B. (2010). Microbial biosynthesis of alkanes. Science 329(5991):559– 562.

Silva D.D.V., Duss′an K.J., Idarraga A., Grangeiro L., Silva S.S., Cardona C.A., et al. (2020). Production and purification of xylitol by *Scheffersomyces amazonenses* via sugarcane hemicellulosic hydrolysate. Biofuels, Bioprod Biorefining,14: 344–56.

Silva-Fernandes T., Santos J.C., Hasmann F., Rodrigues R.C.L.B., Izario Filho H.J., Felipe M.G.A. (2017). Biodegradable alternative for removing toxic compounds from sugarcane bagasse hemicellulosic hydrolysates for valorization in biorefineries. Bioresour Technol, 243:384–92.

Silva D.D. & Felipe M.G.A. (2006). Effect of glucose:xylose ratio on xylose reductase and xylitol dehydrogenase activities from *Candida guilliermondii* in sugarcane bagasse hydrolysate. J Chem Technol Biotechnol;81:1294–300.

Simutis, R. & Lübbert, A. (2015). Bioreactor control improves bioprocess performance. Biotechnol. J, 10, 1115–1130.

Steinwandter, V., Borchert, D. & Herwig, C. (2019). Data science tools and applications on the way to Pharma 4.0. Drug Discov. Today, 24, 1795–1805.

Suttle A. & Sha M. CHO Cell Culture in Eppendorf BioBLU® 10c Single-Use Vessels. Eppendorf Application Note 361. 2018.

Svd H., Härröd M. & Møller P. (1999). Hydrogenation of fatty acid methyl esters to fatty alcohols at supercritical conditions. J Am Oil Chem Soc 76(11):1363–1370

Tabata O., Hashiba K., & Kawakami T. (1993). Inventors; Kao Corporation, assignee. Process for producing alcohol patent US005233099A. 3 August 1993.

Tetala K.K.R. & Vijayalakshmi M.A. (2016). A review on recent developments for biomolecule separation at analytical scale using microfluidic devices. Anal Chim Acta, 906:7-21.

Titchener-Hooker N.J., Dunnill P. & Hoare M. (2008). Micro biochemical engineering to accelerate the design of industrial-scale downstream processes for biopharmaceutical proteins. Biotechnol Bioeng, 100:473-487.

Tornberg E. (2005). Effects of heat on meat proteins – Implications on structure and quality of meat products. Meat Science, 70,493-508.

Tsao C.W. (2016). Polymer microfluidics: simple, low-cost fabrication process bridging academic lab research to commercialized production. Micromachines, 7.

Viner K.J., Champagne P. & Jessop P.G. (2018). Comparison of cell disruption techniques prior to lipid extraction from *Scenedesmus sp.* slurries for biodiesel production using liquid CO2. Green Chem, 20:4330–8.

Wang G., Xiong X., Ghogare R., Wang P., Meng Y. & Chen S. (2016). Exploring fatty alcohol-producing capability of *Yarrowia lipolytica*. Biotechnol Biofuels 9(1):107.

Wewetzer SJ, Kunze M, Ladner T, Luchterhand B, Roth S, Rahmen N, Kloß R, Costa e Silva A, Regestein L, Bu¨chs J: (2015). Parallel use of shake flask and microtiter plate online measuring devices (RAMOS and BioLector) reduces the number of experiments in laboratory-scale stirred tank bioreactors. J Biol Eng, 9:9.

Xin F., Dong W., Zhang W., Ma J. & Jiang M. (2018). Biobutanol production from crystalline cellulose through consolidated bioprocessing. Trends Biotechnol, 37, 167–180.

Xu C., Paone E., Rodríguez-Padr´on D., Luque R., Mauriello F. (2020). Reductive catalytic routes towards sustainable production of hydrogen, fuels and chemicals from biomass derived polyols. Renew Sustain Energy Rev,127,109852.

Xu, S., He, H. & Luo L. (2016a). Status and prospects of municipal solid waste to energy technologies in China. in: O.P. Karthikeyan, K. Heimann, S.S. Muthu, (Eds.), Recycling of Solid Waste for Biofuels and Bio-chemicals. Springer Singapore, Singapore, 31-54.

Xu Y., Xie X., Duan Y., Wang L., Cheng Z., Cheng J. (2016b). A review of impedance measurements of whole cells. Biosens Bioelectron, 77:824-836.

Volk H. & George S.C. (2019). Organic Geochemistry Using petroleum inclusions to trace petroleum systems – a review. Org. Geochem, 129, 99–123.

Yamakawa C.K., Kastell L., Mahler M.R., Martinez J.L., Mussatto S.I. (2020). Exploiting new biorefinery models using non-conventional yeasts and their implications for sustainability. Bioresour Technol, 309:123374.

Yang M., Kuittinen S., Vepsäläinen J., Zhang, J. & Pappinen A. (2017). Enhanced acetone butanol-ethanol production from lignocellulosic hydrolysates by using starchy slurry as supplement. Bioresour. Technol, 243, 126–134.

Yang ST (2007). Bioprocessing – from Biotechnology to Biorefinery: Bioprocessing for Value-Added Products from Renewable Resources. Elsevier B.V., Ch.1, 1-23.

Yee, J.C., Rehmann, M.S., Yao, G., Sowa, S.W., Aron, K.L., Tian, J., Borys, M.C. & Li, Z.J. (2018). Advances in process control strategies for mammalian fed-batch cultures. Curr. Opin. Chem. Eng., 22, 34–41.

Yeong S.K., Idris Z. & Hassan H.A. (2012). Palm oleochemicals in non-food applications, in Palm Oil, ed. by Lai O-M, Tan C-P and Akoh CC. AOCS Press, Washington, DC, 587–624.

Zhou Y.J., Buijs N.A., Zhu Z., Qin J., Siewers V. & Nielsen J. (2016). Production of fatty acid-derived oleochemicals and biofuels by synthetic yeast cell factories. Nat Commun 7(1):11709.

2

Dietary Fibre Concentrates from Fruit by-Products As Promising Source of Prebiotics

Uroosa Noor, Shashi Soni, Ena Gupta

Department of Family and Community Science, University of Allahabad Prayagraj, Uttar Pradesh

Abstract

Dietary fibre concentrates (DFC) are the major component of fruit by-products generated during the processing at commercial level. These are the blessings for human gut health along with many other positive health benefits. Fruit by-products are the so-called nuisance for the environment that enfolds numerous positive signs inside. Lignin, cellulose, hemicelluloses, pectin, waxes, mucilage's, oligosaccharide, gums and many other associated compounds are the part of DFCs. These DFCs when taken in the form of functional ingredient reaches to the colonic microbiota and consequently provides them a healthy growth medium that promotes the cell metabolism upto 67%, enhances the growth of good gut bacteria like Bifidobacterium and Lactobacillus. This article enlightens the content of DFCs in various fruit by-products that shows their prebiotic potential.

Introduction

The term 'dietary fibre' was first coined by Hipsley for non-digestible plant components (hemicelluloses, cellulose, modified cellulose, lignin, mucilage, gums, pectin, oligosaccharides and other associated minor compounds like as cutin, waxes and suberin) present as an integral part of the cell walls of fruits and vegetables, which plays a crucial role on individual health. (Garcia-Amezquita et al., 2018).The new datasets of Recommended Dietary Allowances (RDA) given by expert groupof Indian Council of Medical Research and National Institute of Nutrition (ICMR-NIN), 2020 recommended 30g of dietary fibre per 2000kcal of energy on daily basis (RDA, 2020). Extensive sources of dietary fibre are available in daily meal but due to increased tendency towards modernization and shifting preferences for fast food and junk food leads to diminished consumption of fibre that ultimately imparts bad impression on human gut health. To cope up with this problem many researchers have performed some innovative ideas like supplementing the food with dietary fibre concentrates obtained from fruit by-products, which have been proved to be eco-friendly with best results.

Fruit by-products are composed of higher content of dietary fibre with more varied compositions than cereal and their by-products. Due to their unique applicability and technological functionality, dietary fibre concentrates (DFC) acquired from the by-products of fruit and vegetable processing are most suitable for the formulation of functional food products(Garcia-Amezquita et al., 2018; Dias et al., 2020). DFC provides antioxidant activity, protection against coronary heart diseases and also functioning as rheological modifiers in food (Nieto-Calvache et al., 2019; Patil and Bharati, 2019).

Polysaccharides, majorly fermentable dietary fibres are prebiotics that defend against gastric acid, gastrointestinal absorption and enzymatic hydrolysis. These prebiotics enters the gut microbiota and imparts beneficial effects to the individual's health, for instance short chain fatty acids (SCFA) production, boosting immune power and reducing the population of pathogenic bacteria(Garcia-Amezquita et al., 2018b). Figure 1 illustrates the mechanism of action of fruit by-products on individual's health. This article aims to review the previous published literature on prebiotic potential of fruit by-products generated during the processing of fruits.

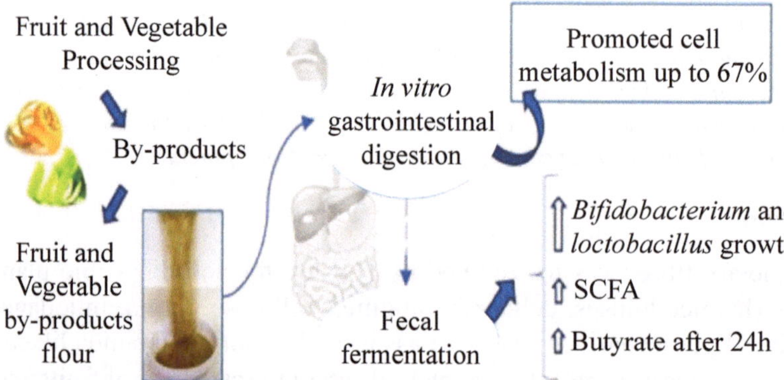

Fig. 1: Mechanism of action of fruit by-products on human body (de Anadrade et al., 2020)

Briefing the dietary fibre concentrates present in fruit by-products

Every fruits and vegetables constitute varied composition of dietary fibre in its cell wall. Although some of them are rich sources of DFC and some are poor sources of DFC. Fruit by-product includes pomace, solid residue peel/

skin, seeds etc. A number of methods are available for extracting the DFC and for its quantification. One of them is alcohol-washing method through which DFCs were revived from agro-industrial by-products namely blackcurrant peel (BCP), banana peel (BP) and clementine peel (CP). Highest soluble dietary fibre (SDF) and lowest insoluble dietary fibre (IDF) content was noted in CP as 34.76 and 42.55g/100g respectively. Whereas, BCP contains highest IDF and lowest SDF content as 68.73 and 8.14g/100g respectively. Total dietary fibre (TDF) content was good enough in all peels as 76.87, 65.55 and 77.32g/100g in BCP, BP and CP, respectively (He et al., 2021).

Slight difference in DFC content was noted when different drying techniques was applied. Freeze-dried (FD) samples exhibits 46.3, 33.4 and 34.0g/100g of IDF along with 7.4, 21.8 and 6.1g/100g of SDF for orange, mango and prickly pear peel, respectively. On the other hand convective hot-air dried (HA) samples contain 42.7, 32.3 and 40.7g/100g of IDF and 6.4, 20.8 and 13.3g/100g of SDF for orange, mango and prickly pear peel, respectively. It is ultimately showing that fruit peels are the mine of DFCs which is the best ever source of prebiotics (Garcia-Amezquita et al., 2018b).

For the production of DFCs from the pulp and peel of papaya, it was treated with ethanol and then dehydrated with two methods viz. microwave (MW) and hot-air convection (CV). Results of chemical analysis showed that the produced DFCs were a combination of lignin, cellulose, non-cellulosic carbohydrates and proteins. Peel of papaya showed far better antioxidative power than the pulp. Overall it has been proved as a promising additive in a diverse range of food products (Nieto-Calvache et al., 2019).

Stenospermocarpic mango (*Mangifera indica* L. cv Ataulfo) is an underutilized fruit that is considered as by-product because of its reduced size and absence of seed. Study on the color, polyphenol content, antioxidant capacity and chemical composition of pulp and peel of stenospermocarpic mango showed that peel flour exhibits higher dietary fibre content that is 40.5g/100g from which 18.19g/100g accounts to SDF and 22.39g/100g is of IDF, than pulp flour (11.7g/100g). It also shows good antioxidative and polyphenolic content so that it can be utilized as an alternative ingredient in the preparation of dietary fibre rich functional food (Patino-Rodriguez et al., 2020).

With the consideration of these statements it can be said that these by-products are not nuisance for environment but are blessings for human beings as well as for ecosystem.

Functional Properties of Dietary Fibre Concentrate

Dietary fibre is mainly present in wide range of vegetative sources like legumes, grains, fruits and vegetables majorly classified as SDF, IDF and TDF. As the name soluble does not mean fibres which are dissolved in water rather it is categorized to provide some physiological functions. However, the ratio of SDF/IDF is crucial for both dietary and functional attributes. Dietary fibre must have the following characteristics such as low caloric value, low moisture and lipid content, high TDF content (more than 50%), bland taste and flavor when used as ingredients in foods.

Dietary fibres are mainly obtained from cereals and grains, which possess abundant SDF and TDF content, low phytic acid and caloric value. Thus, improving the colonic fermentability and functional properties likes water and oil holding capacity. Cereals have more insoluble dietary fibre like cellulose and hemicellulose whereas plant fibre contains more of soluble dietary fibre which holds higher functional property in terms of water holding capacity (WHC), swelling capacity, cation-exchange capacity and viscosity or gel forming capacity that provides additional information about the physiological effect of dietary fibre. These functional attributes of the plant fibre is basically depends on the ratio of SDF/IDF, extraction condition, particle size and source of fibre (vegetable or fruit) (Figuerola, et al., 2005; Garcia-Amezquita, et al., 2018a). At present time, low cost DF sources like fruits by-products has gaining popularity because it provides physiological capacities and increase the technical functionality when used as a dietary ingredient. DFC is obtained by different manufacturing process such as beverage, juice and canning industries which produces large quantity of by-products mainly peel and pomace and are exploited as bulking agent in food to increase the amount of food without increasing the caloric value and cost. Water holding capacity (WHC), swelling capacity (SC), oil holding capacity (OHC), bulk density and water solubility index (WSI) are the major functional properties of DFC which depends on the ratio of SDF/IDF and the chemical composition of DF (Garcia-Amezquita, et al., 2018a). WHC of DFC is due to the fibre content mainly SDF that holds water when come in contact with peripheral centrifugal gravity force. Hydration properties of fibre is because of polysaccharide content along with other parameters like porosity, SDF/IDF ratio, extraction condition, ionic strength, temperature, vegetable source and pressure on the fibre whereas OHC of DFC mainly depends on structure of plant polysaccharides, overall change density, thickness, surface property and hydrophobic nature of the fibre (Serna-Cock, et al., 2016). WHC of DF helps to prevent and treat the intestinal disorders by increasing the bulk of the feaces and decreasing the gastrointestinal transit time. In the food industry, WHC of DF may be used as a functional ingredient to avoid the syneresis and to improve the texture and viscosity of the developed food products, while OHC of DF provides stabilization of fat in the products based on emulsion along with SC of DF increases the viscosity of the food material that is mainly due to SDF content which reduces the absorption of nutrient from the intestinal mucosa and also reduce the postprandial blood glucose and insulin responses. Most important functional property of DF is prebiotic potential. DF helps to ferment the intestinal microbiota that produces short-chain fatty acid (SCFA) such as propionic acid, butyric acid and acetic acid which are linked with beneficial health properties like providing protection against gastro-intestinal disorders, anticancer activity etc. (Rivas et al., 2021).

Prebiotic Potential of Fruit by-products

In current scenario agro-industrial by-products mainly fruit peels and seeds are gaining popularity because of its bioactive compounds like polyphenols and non-phenolic compounds such as unsaturated fatty acid, dietary fibre, prebiotics, bioactive peptides, carotenoids and vitamins that possess a wide variety of beneficial effects on human health without showing any adverse effects as compared to medicinal drugs.By-products from passion fruit, pineapple and pumpkin show the anti-inflammatory, anti-oxidative and anti-dysbiotic effects. It also provides protection against inflammatory bowel diseases (IBDS) mainly crohn's disease and ulcerative colitis through the non-phenolic compounds (Do Nascimento et al., 2021).

Casarottia et al., 2018 investigated the effect of guava, orange and passion fruit by-products incorporated in fermented goat milk and cereal-based fermented products. These fruits by-product contains huge amount of phenolic compounds, carotenoids, antioxidants and fibres that have health-enhancing properties. Fibres present in these fruit peels shows the prebiotic effect that comprises of good alternative for producing probiotic fermented products as value added ingredient with good pro-ecological option.

Akter et al., 2021 and Banerjee et al., 2018 reviewed the prebiotic activity of several fruit peels (grapefruit, pear, pineapple and orange). Pectic oligosaccharides (POS), xylooligosaccharides (XOS) and raffinose (type of resistant starch) were present in all these fruit peels that have the prebiotic potential which improves the host health by enhancing the colon activity, metabolic process, immunomodulatory function and boosts the growth of good bacteria. Beside these activities, XOS also provide protection against pro-carcinogenic enzymes.

Rolim et al., 2018 and 2020 investigated the in vitro prebiotic potential of melon residue (peel and seed) which is considered as lignocellulolitic due to the occurrence of lignin and cellulose accompanied with other compounds like pectin and hemicellulose. These fibres possess a good prebiotic potential with several health benefits like anti-infectious, immunomodulatory activity against free radicals and also as natural additives for food and fibre sources. Banana peel contains good amount of fructooligosaccharides that possesses prebiotic potential that nourishes the gut friendly bacteria. These bacteria producedigestive enzymes and vitamins which improves the ability of human body to better absorb the nutrients (Vu et al., 2018).

Mango peel accommodate high amount of dietary fibres (oligosaccharides and higher structures besides non-digestible compounds) and total soluble sugars or reducing sugars which increases the growth of probiotic bacteria and also promotes the probiotic colonization in gastrointestinal tract with positive health outcomes on the host (Perez-Chabela et al., 2021). In a study three oligosaccharides were obtained by adjusting trichloroacetic acid, in which pectin oligosaccharides (POS) fraction were obtained by controlling the degradation of citrus fruit peel pectin. POS shows the prebiotic effect and many other functional properties by regulating the gut micro flora. Some of the studies proved that oral intake of acidic oligosaccharide derived from pectin helps in the stimulation of intestinal microbiota and production of short chain fatty acid that may leads to the modification of immune responses of chronic lung infection caused by P. aeruginosa (Zhang et al., 2018).

Vital Role of Prebiotics for the Growth of Probiotics

Human digestive enzymes do not digest ample amount of carbohydrate as it extends in the colonic region that is the major source of energy for the colonic microbiota. Peeling and cutting are the most common techniques in fruit processing that generates large amount of waste which may lead to microbial deterioration, these fruit peels and other by-products are the major economical derivatives of dietary fibre and other bioactive phenolic compounds that have the prebiotic potential which is important for the stimulatory effect of gut friendly bacteria from the generaof *Bifidobacterium, Lactobacillus* and *Faecali bacterium prautsnitzii*. These probiotic bacteria possess the potential to biotransform the phenolic compounds into their metabolites, which then regulates the intestinal microbiota, ultimately providing positive health outcomes. Several studies confirmed that addition of phenolic compound in milk based products increases the survival of probiotic culture (Vieira et al., 2017; Sarkar, 2019; Massa et al., 2020). Different fruit by-products such as acerola, orange, mango and passion fruit were used as a prebiotic to detect the growth of probiotic. It was detected that the highest growth of bacterial population of all strains; three commercial starter culture (*Streptococcus thermophiles),* ten probiotic strains (seven *Lactobacillus* spp. and three *Bifidobacterium* spp. strains) and two detrimental pathogenic strain (*Escherichia coli* and *Clostridium perfringens)* was due to the by-product of orange and passion fruit, besides these acerola by-product improves the survival rate of beneficial bacteria while passion fruit by-product stimulates the growth of probiotic bacteria (Vieira et al., 2017).

Vicenssuto et al., in 2020 incorporated mango peel in kefir milk fermentation as a prebiotic to improve the antioxidant and probiotic properties of kefir milk. Results revealed that *Bifidobacterium* a genus of bacteria that produces an enzyme which breakdown the non-digestible portion of mango peel, indicating that it contains a good prebiotic potential and positively increases the growth of lactic acid bacteria. Another study conducted by Massa et al., 2020 examined the prebiotic effect of jabuticaba by-product (JB) and estimated their stimulatory effect on the growth and metabolism of probiotic bacteria from the species *L. acidophilus, L. casei* and *bifidobacterium.* It was found that JB contains high amount of soluble fibres and variety of phenolic compounds which exert the prebiotic property and optimistic modulatory effect on intestinal microbiota.

All the above literature highlights the fact that high DFC content in the fruit by-product act as carrier for the growth of probiotic by positively increasing the number of gut friendly bacteria mainly *Lactic acid bacteria* and *Bifidobacteria.* In addition, it also contains many bioactive compounds that serve many physiological actions such as anti-inflammatory, anti-allergic, anti-carcinogenic, anti-microbial and anti-hypertensive effects.

Applications of Dietary Fiber Concentrate in Food Industry

In this dynamic world the need and fashion of food habit changes from time to time. Due to haphazard eating pattern and lack of time because of workload on general population, it is required to manufacture nutritionally dense food products by applying some innovative technologies like fortification, supplementation, fermentation and many more. To make the food rich in dietary fibre concentrates and also economical, it is suggested to extract the DFCs from fruit and vegetable by-products which can be incorporated in common foods that are often eaten or preferred by the society. For example addition of DFCs in butter improves its bio-functional properties. In this regards wastes generated from orange, orange albedo, celery leaves and root, stone pears and spinach were added in butter and its analysis for proximate value was done. Other parameters such as pH value, color analysis, retardation index, water and oil-holding capacities of the samples were also estimated. Observed ranges for pH values, ash and dry matter was as 3.90-6.48, 0.45-0.96% and 5.79-8.87%, respectively. Technological properties of samples were changed as 65.45-87.32%, 1.8-6.4mL/g and 2.3-8.4mL/g for fiber contents, water and oil holding capacities, respectively. However, reduced glucose in dialysates of all dietary fibre concentrates were compared with control. In butter dietary fibre concentrates were added in different ratios as 3% and 5% to improve the bio-functional properties. Results of color and textural properties of butter showed that sensory evaluation and color parameters has the same effect whereas, type and content of dietary fibre concentrate in the butter changed the textural properties by like hardness and spreadability. Among each groups samples with 5% orange albedo dietary fibre concentrate, 3% stone pear dietary fibre concentrates were mostly preferred in accordance to simple additive weighing techniques (Sarac and Dogan, 2016).

In this way the demand of healthier food can be fulfilled which consists of natural antioxidant power along with dietary fibre. The industrial processing waste of 'Ataulfo' mango peel and paste was utilized in a study where physical, sensory, dietary fiber, polyphenols and antioxidant capacity was analyzed for observing the consequences of using dried 'Ataulfo' mango concentrate (DAMC) and peel and paste of dried 'Ataulfo' mango by-product (DAMB) at 75:25, 50:50, 25:75 ratios for DAMC: DAMB, to replace total sugarcane content in starch-molded mango snacks. Results of the study revealed that dietary fibre content was increased in starch-molded mango snacks from 5.46g to 13.39g/100g dw along with five times increase in polyphenolic content as 107.89 to 626mg GAE/100g dw at 25:75 ratio by replacing sugarcane. Better antioxidant property without change in hardness was noted on replacing all the sugarcane with DAMB in starch-molded mango snacks. Overall, it was suggested that incorporation of DAMC and DAMB in 25:75 ratio gives the best result without affecting its physical and chemical parameters such as color, pH and water activity (Blancas-Benitez et al., 2015).

One of the most consumed tropical food, orange, also produces major amount of by-products from industrial sector. Dietary fibre-rich orange bagasse product (DFROBP) was obtained and its application in the formulation of muffin (a bakery product) with its chemical composition was evaluated. *In vitro* starch digestibility, predicted glyceamic index and acceptability in a sensory test was studied for two different levels of DFROBP added in muffin samples and high dietary fibre contents with low fat content was noted. Effective health beneficial balanced fibre content was analyzed for soluble and insoluble fractions both. Reference and DFROBP contained muffins illustrated similar rapidly digestible starch and at the same time increment in slowly digestible starch was analyzed with the incorporation of DFROBP. Though, the level of resistant starch was reduced with the increased level of DFROBP in the muffins. Predicted glyceamic index was decreased after the incorporation of DFROBP while the two different ratios showed no significant difference. In sensory evaluation, no significant difference was observed in control and in 10% incorporation of DFROBP in muffins. So, it was evident that the addition of DFROBP could be a great alternative for bakery products in food industry especially for those who require low glyceamic response (Romero-Lopez et al., 2011). Overall, the dietary fibre concentrates derived from fruit by-products shows the interesting effect on human health, so are the major fibre sources of food products at industrial level.

Conclusion

This article concludes that fruit by-products are the potential derivatives of non-digestible component of the plant mainly dietary fibres such as hemicelluloses, cellulose, gums, lignin, modified cellulose, mucilage, pectin, oligosaccharides along with other minor compounds and many bioactive phenolic compounds associated with it. Diverse ranges of DFC contents are noted in various fruit by-products. Even though, all fruits by-product contains reasonable amount of DFCs so that it can be included as a functional ingredient in foodstuffs. Subsequently, enhancing its prebiotic potential and health benefits as well as creating an environment friendly relation.

Conflict of Interest Statement

All the authors declare no conflict of interest. There is no financial interest to report and all co-authors have seen and agree with the contents of the manuscript.

References

Akter, B., & Rabeta, M. S. (2021). Synbiotic and antioxidant activity of fruit by-products and their effect on human health. Food Research, 5(1), 24-35. https://doi.org/10.26656/fr.2017.5(1).401

Banerjee, S., Ranganathan, V., Patti, A., &Arora, A. (2018). Valorisation of pineapple wastes for food and therapeutic applications. Trends in food science & technology, 82, 60-70. https://doi.org/10.1016/j.tifs.2018.09.024

Blancas-Benitez, F. J., de JesúsAvena-Bustillos, R., Montalvo-González, E., Sáyago-Ayerdi, S. G., & McHugh, T. H. (2015). Addition of dried 'Ataulfo'mango (Mangiferaindica L) by-products as a source of dietary fiber and polyphenols in starch molded mango snacks. Journal of Food Science and Technology, 52(11), 7393-7400.

Casarotti, S. N., Borgonovi, T. F., Batista, C. L., & Penna, A. L. B. (2018). Guava, orange and passion fruit by-products: Characterization and its impacts on kinetics of acidification and properties of probiotic fermented products. LWT, 98, 69-76. https://doi.org/10.1016/j.lwt.2018.08.010

de Andrade, R. M. S., Silva, S., Costa, C. M. D. S. F., Veiga, M., Costa, E., Ferreira, M. S. L., ... & Pintado, M. E. (2020). Potential prebiotic effect of fruit and vegetable byproducts flour using in vitro gastrointestinal digestion. Food Research International, 137, 109354.

Dias, P. G. I., Sajiwanie, J. W. A., & Rathnayaka, R. M. U. S. K. (2020). Chemical composition, physicochemical and technological properties of selected fruit peels as a potential food source. International Journal of Fruit Science, 20 (sup2), S240-S251. DOI: 10.1080/15538362.2020.1717402.

Do Nascimento, R. D. P., Moya, A. M. T. M., da Fonseca Machado, A. P., Geraldi, M. V., Diez-Echave, P., Vezza, T., ... & Junior, M. R. M. (2021). Review on the potential application of non-phenolic compounds from native Latin American food byproducts in inflammatory bowel diseases. Food Research International, 139, 109796. https://doi.org/10.1016/j.foodres.2020.109796

Figuerola, F., Hurtado, M. L., Estévez, A. M., Chiffelle, I., & Asenjo, F. (2005). Fibre concentrates from apple pomace and citrus peel as potential fibre sources for food enrichment. Food Chemistry, 91(3), 395-401.

Garcia-Amezquita, L. E., Tejada-Ortigoza, V., Serna-Saldivar, S. O., & Welti-Chanes, J. (2018a). Dietary fiber concentrates from fruit and vegetable by-products: processing, modification, and application as functional ingredients. Food and Bioprocess Technology, 11(8), 1439-1463. https://doi.org/10.1007/s11947-018-2117-2.

Garcia-Amezquita, L. E., Tejada-Ortigoza, V., Campanella, O. H., &Welti-Chanes, J. (2018b). Influence of drying method on the composition, physicochemical properties, and prebiotic potential of dietary fibre concentrates from fruit peels. Journal of Food Quality, 2018. https://doi.org/10.1155/2018/910523.

He, C., Sampers, I., & Raes, K. (2021). Dietary fiber concentrates recovered from agro-industrial by-products: Functional properties and application as physical carriers for probiotics. Food Hydrocolloids, 111, 106175.

Massa, N. M. L., Menezes, F. N. D. D., de Albuquerque, T. M. R., de Oliveira, S. P. A., dos Santos Lima, M., Magnani, M., & de Souza, E. L. (2020). Effects of digested jabuticaba (Myrciariajaboticaba (Vell.) Berg) by-product on growth and metabolism of Lactobacillus and Bifidobacterium indicate prebiotic properties. LWT, 131, 109766. https://doi.org/10.1016/j.lwt.2020.109766

Nieto-Calvache, J. E., de EscaladaPla, M., &Gerschenson, L. N. (2019). Dietary fibre concentrates produced from papaya by-products for agroindustrial waste valorisation. International Journal of Food Science & Technology, 54(4), 1074-1080.

Patil A V and Bharati P. (2019). Antioxidant dietary fibre: A healthy functional ingredient. The Pharma Innovative Journal, 8(9), 272-275.

Patiño-Rodríguez, O., Bello-Pérez, L. A., Agama-Acevedo, E., & Pacheco-Vargas, G. (2020). Pulp and peel of unripe stenospermocarpic mango (Mangiferaindica L. cvAtaulfo) as an alternative source of starch, polyphenols and dietary fibre. Food Research International, 138, 109719.

Pérez-Chabela, M. D.L., Cebollón-Juárez, A., Bosquez-Molina, E., & Totosaus, A. (2021). Mango peel flour and potato peel flour as bioactive ingredients in the formulation of functional yogurt. Food Science and Technology. https://doi.org/10.1590/fst.38220

RDA Short Report_Dec 09 2020: Retrived from https://drive.google.com/file/d/1j3umH5zcJAGNR_WUFwl3 0rBiemYw8DR/view?usp=embed_facebook (last checked on May. 6th 2021).

Rivas, M. Á., Casquete, R., Córdoba, M. D. G., Ruíz-Moyano, S., Benito, M. J., Pérez-Nevado, F., & Martín, A. (2021). Chemical Composition and Functional Properties of Dietary Fibre Concentrates from Winemaking By-Products: Skins, Stems and Lees. Foods, 10(7), 1510.

Rolim P M, de Oliveira Junior S D, de Oliveira A N A, Mendes C, dos Santos E S and deMacedo G R. (2018). Nutritional value, cellulase activity and prebiotic effect of melon residues (*Cucumismelo* L. reticulatus group) as a fermentative substrate. J. Food Nutr. Res,57(4), 315-327.

Rolim, P. M., Seabra, L. M. A. J., & de Macedo, G. R. (2020). Melon by-products: Biopotential in human health and food processing. Food Reviews International, 36(1), 15-38. https://doi.org/10.1080/87559129.2019.1613662

Romero-Lopez, M. R., Osorio-Diaz, P., Bello-Perez, L. A., Tovar, J., & Bernardino-Nicanor, A. (2011). Fiber concentrate from orange (*Citrus sinensis* L.) bagase: characterization and application as bakery product ingredient. International Journal of Molecular Sciences, 12(4), 2174-2186.

Saraç, M. G., &Dogan, M. (2016). Incorporation of dietary fiber concentrates from fruit and vegetable wastes in butter: effects on physicochemical, textural, and sensory properties. European Food Research and Technology, 242(8), 1331-1342.

Sarkar, S. (2019). Potentiality of Probiotic Fruit Yoghurt as a Functional Food–A Review. J Nutr Food Sci, 2, 100009.

Serna-Cock, L., García-Gonzales, E., & Torres-León, C. (2016). Agro-industrial potential of the mango peel based on its nutritional and functional properties. Food Reviews International, 32(4), 364-376.

Vicenssuto, G. M., & de Castro, R. J. S. (2020). Development of a novel probiotic milk product with enhanced antioxidant properties using mango peel as a fermentation substrate. Biocatalysis and Agricultural Biotechnology, 24, 101564. https://doi.org/10.1016/j.bcab.2020.101564.

Vieira, A. D. S., Bedani, R., Albuquerque, M. A. C., Biscola, V., &Saad, S. M. I. (2017). The impact of fruit and soybean by-products and amaranth on the growth of probiotic and starter microorganisms. Food Research International, 97, 356-363. https://doi.org/10.1016/j.foodres.2017.04.026

Vu, H. T., Scarlett, C. J., &Vuong, Q. V. (2018). Phenolic compounds within banana peel and their potential uses: A review. Journal of Functional Foods, 40, 238-248. https://doi.org/10.1016/j.jff.2017.11.006

Zhang, S., Hu, H., Wang, L., Liu, F., & Pan, S. (2018). Preparation and prebiotic potential of pectin oligosaccharides obtained from citrus peel pectin. Food chemistry, 244, 232-237. https://doi.org/10.1016/j.foodchem.2017.10.071

[illegible] nutritional value [illegible] activity [illegible] (or [illegible] group) [illegible] Journal of Food [illegible]

[illegible] (2012) [illegible] by-products [illegible] 13-24 [illegible]

[illegible] characterization and [illegible] Journal of [illegible]

[illegible] (2019). Incorporation of dietary fibre concentrates from fruit and vegetable wastes in [illegible] effects on physicochemical, textural and sensory properties. International Journal of Food Science and Technology, 54(4): 1351-1[illegible]

[illegible] (2019). Potentiality of [illegible] as a functional [illegible] Food Sci. [illegible]

[illegible] (2019). [illegible] products. Food [illegible]

[illegible] (2019). Development of a novel probiotic milk product with enhanced [illegible]

[illegible] important role [illegible] analysis of [illegible] and [illegible] Food Research International [illegible]

[illegible] compounds [illegible] and their potential [illegible] Journal of Functional Foods, 40: [illegible]

[illegible]

3

Development of Starch-Based Bioproducts

Neelam Yadav and Neelam Dubey

Centre of Food Technology, Science Faculty Campus
University of Allahabad, Prayagraj, Uttar Pradesh

Abstract

Starch is an important plant-based product and an abundant biomaterial, used worldwide for different purposes in many industrial sectors including food, health, chemical and engineering sector. Starch versatility in industrial applications is largely defined by its physicochemical properties and functionality. Starch in its native form has limited functionality and application but advancements in technological processing have led to the wide-range modification of starch for different purposes. It can be easily degraded by microorganisms hence is a natural source of bioenergy in the form of bioethanol and biohydrogen as well as in the development of biodegradable packaging materials. Various technological interventions can significantly improve the characteristics of native starch and be tailor-made to the specific needs of the packaging and food industry. Native starch can be treated by various methods to modify its characteristics. Modification improves starch paste clarity, viscosity, syneresis and freeze-thaw stability leading to a wide range of food applications. Esterification lowers gelatinization temperature and retrogradation lowers the tendency to form gel and higher paste clarity which shows applications in refrigerated and frozen foods, as emulsion stabilizers and for encapsulation. Cross-linking increases the stability of granules towards swelling, high temperature, high shear, and acidic conditions. Starch-based nano-composite polymers have their application in coating and packaging of food as well as non food commodities. Techniques involving extrusion, casting, spinning, and nanotechnology can be used in the production of films, foams, and nanocomposite materials. Due to the unique characteristics of resistant starch, it is being used as a valuable supplement in the formulation of drugs, pharmaceuticals, various functional foods and in the microencapsulation of probiotics and bioactive compounds.

Introduction

Starch is the main source of energy in the food chain where light energy is converted into chemical energy and all the living organisms are directly or indirectly dependent on it. It is also the stable form of energy in the diet of human being as approximately 40-80% of energy comes from the starch or starch-based product in our daily diet. Besides food starch and its biomolecules have got a wide application in paper, packaging, adhesive, textile, cosmetic, and pharmaceutical industry and is also used in bioplastic production. Its applications in the food industry is wide and diverse and are in high demand in the manufacturing of frozen, ready to eat convenience food as well as packaged food. Its industrial application as a thickener, stabilizer, binder, and gelling agent is expected to raise the starch market in the coming years at a growth rate of 7.0% (compound annual growth rate, CAGR) in the world during 2020-2028. The current global market is 97.85 billion USD (Global Industrial Starch Market Size Report).

Sources of Starch

Cereals, pulses, roots, and tubers are the major sources starch from plants. In grains, it is mainly stored in the endosperm in different sizes and shapes according to species whereas in roots and tubers it is present in the parenchyma. Among grains, wheat, rice, corn, barley and in roots and tubers potato, cassava, and sweet potato are the major sources of starch. Starch content ranges between 40-90% in cereals, 20-50% in legumes, 30-70% in roots, and 65-85% in tubers on dry matter basis. In addition to this raw fruits like banana, mango (approximately 70% on dry weight basis) are also appreciable sources of starch. Seeds of jackfruit also contain 60-80% starch on a dry weight basis. Apart from this, stem system of plant *Metroxylon sagu* known as a sago plant, is also a commercial source of starch (Karim et al., 2008). Some novel sources of starch are Ramon (*Brosimumalicastrum*) seeds containing 63% starch (Olguin et al., 2017). Another novel source of starch is microalgae like Chlorophyceae containing 40% starch on dry weight basis (Gifuni et al., 2017). Hence, starch can also be explored as an important source of bioproducts. Bioproducts are materials made from renewable or biological materials and often offer the opportunity to farmers and food processors for economic gain. These products reduce our dependency on non-renewable resources and produce less harm to the environment. Starch-based bioproducts are categorized into bioenergy, biomaterials, and various biochemicals.

Starch-based Bio-energy

Limited availability of crude oil, non renewable energy sources and increasing concern towards environmental problems has compelled us to think towards alternative resources of energy (Demirbas, 2016). Plant based carbohydrate substrate can be bioprocessed for using it as an alternative source of energy and will be more cost effective as it reduce the combustion pollutant of fossil fuel (CO_2, NO_x, SO_2) (Francielo, 2015). Starch is present in large quantity throughout the globe and its biomass can be a source of energy in the form of biohydrogen, bioethanol and other liquid fuels.

Biohydrogen

Fermentation of carbohydrate (starch, cellulose) under normal environmental temperature and presence can be used for production of hydrogen. It is much easier and cost effective method in comparison to thermochemical and electrochemical method and at the same time it is environmental friendly with zero carbon emission. Biohydrogen has higher energy content per weight unit (122 kJ g^{-1}) and energy yield. Its energy content is about 2.8 times higher than gasoline and diesel (Nogueira, et al., 2015). Aerobic, anaerobic and photosynthetic bacteria and algae are capable of biohydrogen production; however fermentative anaerobic bacteria are most suitable due to high rate of production from organic resources. Hydrogen is produced through fermentation process when anaerobic bacteria are allowed to grow in dark on carbohydrate substrate and acetic acid butyrate is produced as end product as shown in equation (1) & (2).

$$C_6H_{12}O_6 + 2H_2O\ 2CH_3COOH + 4H_2 + 2CO_2 (1)$$

$$C_6H_{12}O_6 + 2H_2O\ CH_2CH_2CH_2COOH + 2H_2 + 2CO_2 (2)$$

Usually Enterobacter, Bacillus and Clostridium species are capable of fermenting carbohydrate rich substrate in dark conditions. Starch in the form of polymer yields different amount of hydrogen per mole of glucose depending upon pathway and its end products.

Bio reactors used for fermentation and hydrogen production may be of different kinds. Research on production of biohydrogen has been increased high in past few decades and different kinds of substrates have been used for production of biohydrogen. Cellulose and starch are the major substrate utilized for biohydrogen production, however starch is more efficient as it can be hydrolyzed through acidic or enzymatic route, producing sugars and dextrins which further acted upon by micro-organisms for hydrogen production. Whereas cellulose requires pre fermentation with cellulolytic enzymes. Byproducts of food industries like biomass of wheat, cassava, corn, sorghum and effluents of potato have been successfully utilized for biohydrogen product from their starches (Fang et al. 2006; Cappelletti et al. 2009, Hussy et al. 2003). Biomass of agricultural waste, residues of brewery, bread and sago industries as well as starch waste water have also been used for this purpose and can be excellent method of waste treatment. Recently application of magnetic nanoparticles has been explored in improving biohydrogen production. Carbon metabolic and non metabolic nanoparticles give a longer surface area and prevent elimination of enzymes leading to higher yield of biofuels (Vasantha et al. 2021).

Liquid Fuels (Ethanol)

Starch plants can be fermented to produce ethanol. Additionally, biological feedstocks contain appreciable amount of sugar or materials that can be converted into sugar, such as starch or cellulose that can be fermented to produce bioethanol (Balat, 2008; Mojovic et al., 2006) which can be further used as an alternative fuel. Bioethanol feedstocks can be conveniently classified into three types: i) sucrose-containing feedstock (e.g., molasses, sugar beet, sweet sorghum, and sugar cane), ii) starch-containing feedstock (e.g., wheat, corn, rice, triticale potato, and barley) and iii) lignocellulosic biomass (e.g., wood, straw, and grasses).

Molasses is a traditional raw material for the production of ethanol. It is obtained as a by-product in the production of sugar from sugar beet (beet molasses) or sugarcane (cane molasses). For a long time, molasses was practically the only raw material for the production of ethanol.

Combustion of ethanol yields a large quantity of energy that can be used to power cars and generate electricity. Since starch plants can be harvested to produce ethanol, the supply of ethanol is renewable during the life cycle of the plant. Fermentation is a natural biochemical process involving the decomposition of starches producing ethanol as a by-product in the absence of oxygen. The process of fermentation can be enhanced by using yeast and heating the vegetation. Nowadays, the focus is on bioethanol production from crops, such as corn, wheat, and sugar cane, as well as on highly abundant agricultural wastes.

Ethanol can be blended with petrol or used as neat alcohol in dedicated engines, taking advantage of the higher-octane number, low cetane number, and higher heat of vaporization and also it is an excellent fuel for future advanced Flexi-fuel hybrid vehicles. Using ethanol fuel (E85 – with 85% of bioethanol) in a midsize passenger vehicle can reduce greenhouse gas emissions by 41–61% km^{-1} driven, compared to gasoline-fueled vehicles (BiELACZYC, et al., 2016).

In addition, ethanol is an oxygenated fuel containing 35% oxygen, which reduces particulate and NOx emissions from combustion; it is biodegradable and contributes to sustainability (BiELACZYC, et al., 2016).

Bioethanol has already been introduced on a large scale in Brazil, US, and European countries (Balat, 2008, Balat et al., 2009; Sanchez and Cardona, 2007; Cardona, 2007; Das Neves et al., 2007; Mojovic et al., 2006; BiELACZYC, et al., 2016). Production of bioethanol has increased remarkably because many countries look for reducing oil imports, boosting rural economies, and improving air quality. The world ethyl alcohol production has reached about 27310 million gallons in 2021 (Statista 2022). USA and Brazil being the first producers they together account for about 70% of the world bioethanol production. On average, 73% of produced ethanol worldwide corresponds to fuel ethanol, 17% to beverage ethanol and 10% to industrial ethanol.

Starch-based Biomaterials

Structural Changes in Starch for the Development of Biomaterials

Starch is a biopolymer of D-glucose made up of amylose and amylopectin and one of the major polysaccharides of carbohydrate present in food. Amylose is a straight chain having molecular weight closer to a synthetic polymer such as polyethylene (PE), polypropylene (PP), polystyrene (PS) in comparison to amylopectin which has branched-chain with higher molecular weight than conventional synthetic polymers (Jiang et al., 2020). These starch granules are usually stored in cereal and tubers having specific shapes and structures ranging from 2-100 μm in size. The utilization of starch for the development of biomaterials like bioplastics, bio foams or other nanocomposites is limited in its native forms. Native starch has a restricted solubility hence it inhibits its industrial use. Starch is either modified or blended with natural polymer to increase its industrial application. Modification of starch is constantly increasing as it improves the properties and emulsifying utility of starch making it suitable for film formation and excellent coating material. This can be a very good biodegradable polysaccharide material by the action of various microorganisms as it is a natural polysaccharide (Jiang et al., 2020).

Native starch due to its polyhydroxy group in the structure is hydrophobic and semi-crystalline. The hydrogen bond between the hydroxyl groups of a starch molecule is the limiting situation for plastic behavior of native starch secondly the branched chain consisting of amylopectin is in the form of double-helical with a length of 5nm and is mainly responsible for the crystalline region (Jiang et al., 2020) leading to nano plastic characteristics. Thermal processing of starch shows special features leading to water diffusion, granule expansion, gelatinization, decomposition, and crystallization. Functional properties of starch (hardness, transparency, and reactivity) are affected by crystallinity. Due to the high molecular weight of amylopectin thermal processing requires multiple reactions, of which gelatinization is the most important reaction. Formation of thermoplastic starch (TPS) takes place so after disruption of starch granules, which is lower than its melting temperature. In gelatinization, granules swell and the crystalline structure of starch granules is destroyed. This can be either recrystallized or retrogradation takes place. So TPS is the conversion of native starch into gelatinized starch-based material in presence of water or plasticizer blending. TPS can further be treated and has application in packaging materials (Kong, 2012).

Extraction and Modification of Starch

Extraction of starch is done mainly by distilled water method or alkaline extraction method. Distilled water method of extraction involves homogenization of the sample with distilled water and passing the slurry through the filter and allowing the filtrate for 3 hours at ambient temperature. The precipitated starch is washed 3-4 times and dried at 50 degrees centigrade and ground to a fine powder (Babu, 2014). The alkaline extraction method produces high purity starch. The process involves a total of 6 steps, which are as follows: (1) mixing flour with alkaline solution (2) protein digestion, (3) primary filtration, (4) water washing, (5) vacuum filtration and (6) drying the starch cake (Usman, 2014).

Modification of Starch

Various environmental, operational and processing factors have caused changes in the structure of starch molecules and are called modifications. Starch is rarely consumed in its intact form and industry often uses it in its natural form for further processing. Most native starches have limited direct application because they are unstable with respect to changes in temperature, pH, and shear forces. Native starches are often modified to develop specific properties such as solubility, texture, adhesion and tolerance to high temperatures used in industrial processes (Sweedman et al., 2013). These modifications can utilize either positive or negative effects on the structure and functionality of the starch molecules. Several methods (physical, chemical and enzymatic modifications) have been developed to produce modified starches with different properties and applications (Boukhalfa et al., 2018; Korma et al., 2016; Nabeshima et al., 2001; Singh et al., 2000; Kulp, 1981; Zia-ud-Din et al., 2017). All these techniques alter the starch polymer, making it highly flexible and altering its physicochemical properties and structural attributes to increase its value to the food and non-food industries (Krithika and Ratnamala, 2019).

Physical Modification

Physical modifications include changes in the morphology and three-dimensional structure of starch under the influence of some physical factors, such as milling, moisture, temperature, pressure, pH, radiation, pulse-electric field, ultrasonic waves, etc (Nawaz et al., 2013). It results in changes in the functional and physico-chemical properties of starch, such as particle size, surface properties, solubility index, and functional properties such as water absorption, swelling capacity, and gelation capacity.

Several studies of physical modification of starch using different techniques have been reported. Commonly used physical modification methods include moisture heat treatment, annealing, retrogradation, starch super heating, gelatinization, heat inhibition treatment, ultra-high pressure, UV and gamma irradiation, microwave treatment, osmotic pressure and instantaneous controlled pressure, mechanical activation by stirring ball mill, pulsed electric field treatment, vacuum ball mill micronization and freeze-thaw treatment (Rosa and Dias, 2011; Huang et al., 2007; Nawaz et al., 2018; Shaoxiao and Baodong, 2013; Lawal et al., 2019). The effects of physical treatment on starch structure and properties are summarized in Table 1.

Table 1: Effects of physical treatment on starch structure and properties

Modification method	Treatment	Change in starch structure	Change in starch properties	References
Gentle heating	Heating starch at a low temperature (45-65 °C)	It causes slight changes in the starch structure and the ratio of amylopectin to amylose.	It does not show any significant effect on the physico-mechanical properties of starch.	Wicaksono et al.(2016)
Hydrothermal treatment	Heating starch in an aqueous environment.	Causes physical reorganization of starch granules.	It improves granule size, mobility and stability, allowing them to be easily digested by amylase. It also improves the gelatinizing properties of starch.	Loisel et al., (2006), Tester et al., (2000)
Heat-moisture treatment	Application of heat in the presence of a limited level of humidity: 22-27% and a high temperature above the glass transition temperature: 100-120°C for a specified time: 1-24 h	The result is a change in the size, shape and granular and crystalline structure of the starch. It also causes disruption of the helical structures in the amorphous regions of the starch granules. It induces the molecular degradation of starch and increases the degree of polarization.	It reduces amylose leaching, maximum viscosity and swelling capacity and increases solubility, thermal stability, gelation temperatures, pasting temperature, pasting time, interaction properties and susceptibility of starch to chemical and enzymatic attack (α-amylase and acid hydrolysis).	Kulp and Lorenz, 1981, BeMiller and Huber, 2015
Ultraviolet (UV) irradiation	Starch granules are exposed to UV radiation.	It causes free radical induced crosslinking and depolymerization, oxidative photodegradation and dextrinization in starch.	UV treatment affects the physical, chemical and functional properties of starch.	Fiedorowicz et al.,(1999); Bertolini et al., (2001); Merlin et al., (1981)
Gamma irradiation	Exposure of starch granules to different doses of high-energy gamma radiation.	Gamma radiation causes the amylopectin chains to break in the amorphous regions and reduces the ratio of amylopectin to amylose. It also causes radiolysis and radiodepolymerization of starch.	Exposure to gamma radiation decreases paste viscosity and starch enthalpy change and amylopectin molecular weight and radius of gyration. It increases the susceptibility of starch to	Bao et al., (2005); Michel et al., (1980); Raffi et al., (1981).

Modification method	Treatment	Change in starch structure	Change in starch properties	References
			amylase. It also improved rheological properties such as gelatinization viscosity, swelling ability and solubility.	
Micronization in a vacuum ball mill		It damages type B starch granules, resulting in loss of grain order and double helix content and reduction of crystallinity.	It also changes the rheological properties of starch. Increases water adsorption, iodine binding capacity, granule swelling and solubility.	BeMiller et al.,(2015); Tamaki et al., (1997); Tamaki et al., (1998).
Ultra-high-pressure treatment	Treatment of starch under pressure > 400 MPa.	It deforms in the crystalline region and transfers type A to type B crystalline starch.	It increases the swelling of starch granules and limits the leaching of amylose. It lowers the gelation temperature.	Wicaksono et al. (2016); Katopo et al., (2002).
Annealing	Modification of starch in the presence of moderate water content (40-50% w/w) or excess water greater than 65% w/w at temperatures below the onset of gelatinization.	It increases the interaction between amylose-amylopectin and amylose-amylose chains and crystalline perfection. It increases the mobility of the segments of the double helix chain in the granules, allowing subsequent recrystallization, restructuring, improves molecular order and provides greater homogeneity between crystallites.	It reduces the leaching and swelling of amylose granules and increases the thermal stability of gelatinization temperatures and susceptibility to α-amylase.	BeMiller et al.,(2015); Omoregie et al.,(2015).
Freeze and thaw treatment	Starch was heated to a high temperature (59–79 °C) and then followed by freezing and thawing.	Increasing the number of free melting cycles changes the complex modulus and phase angle of starch.	It affects the rheological properties of starch. It increases the swelling ability, viscosity and thermal stability of starch. It also affects the surface properties of starch granules.	Szymońska et al., (2005); Eliasson et al, (2008).

Chemical Modification

Chemical modification involves the introduction of functional groups into the starch molecule (without any physical change in the shape and size of the molecule), resulting in significantly altered physicochemical properties. Each of the glucose units in amylose and amylopectin has three reactive hydroxyl groups, which are the main sites for chemical modification in starch (Nawaz et al., 2013). Modification of native granular starches profoundly alters their behaviour during gelatinization, salting, pasting and retrogradation. In addition, chemically modified starch is widely used in the food industry due to its better emulsifying and pasting properties (Wang et al., 2016).

Modification is generally achieved by derivatization, such as etherification (adding some hydroxyethyl, hydroxypropyl or carboxymethyl groups to starch hydroxyl groups), esterification (condensation of some fatty acids, other carboxylic acids and phosphates with active starch hydroxyl groups), cationization by introducing some cationic molecules, cross-linking and graft polymerization of starch with synthetic polymers, degradation (acidic or enzymatic hydrolysis and oxidation of starch) [Greenwood et al., 1977; Hernandez, 2018; Jyothi et al., 2006; Waliszewski et al., 2003; Cizova et al., 2008; Liu et al., 2015; Paleos et al., 2017; Zanella et al., 2019]. However, these techniques are limited due to customer and environmental concerns.

Cationic modifications involve the reaction of starch molecules that contain tertiary and secondary ammonium, imino, amino, sulfur, and phosphate groups that react with starch hydroxyl groups. It improves the dielectric constant of starch granules. It is of great importance in the textile industry as an additive and in the paper and cosmetic industries due to its low price, rapid degradation and bioavailability (Zhang, 2001). Table 2 comprehensively summarizes the effect of some commonly used chemical treatments on the properties of starch.

Table 2: Effects of chemical treatment on starch structure and properties

Modification method	Treatment	Change in starch structure	Change in starch properties	Reference
Etherification				
Hydroxyethylation	Introduction of a hydroxyethyl group into starch.	Changes in granular structure.	It improves the binding capacity of some anticancer and other drugs.	Paleos et al., (2017)
Hydroxypropylation	Introduction of a hydroxypropyl group into starch	It disrupts inter- and intramolecular hydrogen bonds and weakens the granular structure of starch. Addition of a hydroxypropyl group to starch. It increases the emotional freedom of starch chains in amorphous regions.	It increases the maximum viscosity, water-binding capacity, swelling capacity, solubility and enzymatic digestibility of starch. It also improves paste clarity and freeze-thaw stability. However, it reduces the gelatinization parameters.	Liu et al., (1999);
Esterification				
Acetylation	Reaction of the acetyl group with the hydroxyl group of polymeric starch.	Slows down crystallization or retrogradation in starch granules.	It inhibits the formation of intramolecular hydrogen bonds and increases the viscosity and swelling capacity of granules. It reduces the temperature and solubility of the paste.	Betancur et al.,(1997); Thirumdas et al.,(2017).
Phosphorylation	Addition of a phosphate group to the hydroxyl groups of starch.	The result is the formation of monophosphate or diphosphate starch. It increases steric hindrance and prevents the linearity of molecular chains.	Improves the viscosity, textural properties, paste clarity and stability of Freeze Thaw starch. It also increases resistance to low pH, high temperature and high shear. It lowers the gelation temperature.	Waliszewski et al., (2003).
Cationization	Modification of starch with different cationic molecules.	Introduction of amino, ammonium, imino, phosphonium or sulfonium groups to impart a positive ionic charge to the starch.	It increases the solubility, stability, dispersibility and clarity of starch.	Korma et al, (2016).

Modification method	Treatment	Change in starch structure	Change in starch properties	Reference
Dry cationization	Dry heating of citric acid to anhydride.	Cationic molecules are sprayed in the absence of a liquid phase onto the dried starch during extrusion.	The result is the formation of cross-linked starch citrate. It changes the granular structure of starch and improves its adsorption properties.	Liu et al., (2015).
Wet cationization	Homogeneous and heterogeneous reaction of starch with cationic molecules in the presence of a liquid medium.	The result is also the formation of cross-linked starch.	It increases the viscosity and lowers the temperature of the paste.	Liu et al., (2015)
Cross-linking	(Formation of inter- and intramolecular bridges) Etherification and esterification of granules with crosslinking polymers by reaction with a mixture (99:1) of sodium trimetaphosphate and sodium tripolyphosphate or other crosslinking agents in an aqueous alkaline suspension containing sodium sulfate..	It reduces the mobility of amorphous chains in starch granules. It introduces inter- and intramolecular bonds with multifunctional small molecules with hydroxyl groups on starch to strengthen granules against various factors.	It increases the arrangement of the internal structure of the granules and stability. It reduces the solubility of starch in water, which reduces its association with lipids, moisture and proteins. It also causes a decrease in viscosity, swellability, digestibility, retrogradation rate, maximum relaxation endotherm temperature, and starch enthalpy. Increases gelatinization temperature, glass transition temperature, melting enthalpy, free volume of starch chains, relaxation enthalpy and stability of starch at high temperatures.	Boukhalfa et al., 2018, Jyothi et al., 2006

Modification method	Treatment	Change in starch structure	Change in starch properties	Reference
Graft copolymerization	Copolymerization of starch with synthetic polymers such as poly(ethylene terephthalate), polyethylene, polypropylene, polyvinyl chloride and polystyrene	It changes the structure of starch from a homopolymer to a heteropolymer.	Physical properties and reactivity of the starch changes.	Hernández et al.,(2018)

Enzymatic Modification

Enzymatic modification of starch involves exposure of starch suspensions to various enzymes, especially hydrolyzing enzymes, to produce highly functional derivatives. This technique dates back to the days when glucose syrup or high fructose corn syrup was made. Amylomaltase enzymes which are found in representatives of eukarya, bacteria and archaea, break the α-1,4 bond between two glucose units and subsequently create a new α-1,4 bond modified starch that can be used in food, cosmetics, pharmaceuticals, detergents, adhesives and drilling fluids (Krithika et.al., 2019). It is also a good source of vegetable substitute for gelatin except that it forms a cloudy gel whereas gelatin gels are transparent (Kaper et al., 2003).

Hansen et al. (2008) reported that modification of potato, high-amylose potato, corn, and pea starches with amylomaltase (AM) isolated from a hyperthermophilic bacterium, i.e., *Thermus thermophilus*, improved the gel texture compared to the parent starch. In addition, extended amylopectin chain length profiles were observed for all modified starches. Cyclomaltodextrinase (CDase; EC 3.2.1.54), isolated from alkalophilic *Bacillus* sp. *I-5* (CDase I-5) has been used to modify rice starch to produce low-amylose starch products (Krithika and Ratnamala, 2019).

Similarly, treatment of corn starch with β-amylase, β-amylase & transglucosidase, maltogenic α-amylase, and maltogenic α-amylase & transglucosidase resulted in significant decreases in digestion rate by 14.5%, 29.0%, 19.8%, and 31.8 %, or producing resistant starch with a reduced glycemic index, which can be used in diabetes, prediabetes, cardiovascular disease and obesity. An increase in starch branch density and crystalline structure in modified starches is thought to contribute to slow digestion.

Cyclomaltodextrins are formed in situ, with CD retention in the granules, leading to the production of a new material that has the properties of starch granules and cyclomaltodextrins. Complexation of organic molecules with cyclomaltodextrins provides stabilization of light, heat and oxygen sensitive materials in starch granules as well as a mechanism for their slow release, in addition to imparting special tastes, aromas and flavors to starch granules (Krithika and Ratnamala, 2019).

Technologies for Starch-based Bioproducts for Packaging

Various techniques applied for the development of starch-based biomaterials include exclusion, foaming, processing, film coating, and flow methods. Recently electrospinning, force spinning, 3D printing, and reaction extrusions are the new technologies being used for the development of starch and natural materials.

Extrusion- This process is the most commonly used method where gelatinization takes place and there is partial depolymerization leading to a decrease in its crystallinity. The process usually consists of extrusion blowing, compression molding and injection molding.

Foaming-Extrusion processing is used for the foaming process of polymers. In addition, thermal processing, microwave heating, solvent exclusion, and supercritical fluid extraction methods are also used. Foam formation takes place by nucleation process in a gas-saturated polymer at high temperatures (Mittal, 2014). Usually, blends of synthetic fiber or plasticizer are used with starch to improve the mechanical properties of the foam.

Film coating- It consists of pouring the polymers and solution on a flat greased surface followed by drying under controlled temperature and humidity. To provide stiffness to starch-based films usually acetic acid and lactic acid are added to provide stiffness (Garcia-Guzman et al., 2022).

Electrospinning- It is used for the production of nanofibre by electrospinning. The equipment is consisting of electrospray and spinning where polymer solution experiences electrostatic force and Coulombic force. Force spinning in a newer method using centrifugal force with large-scale production of nanofibers.

3D printing- It is a prototyping through three printers, adding starch and 2 to 5% glycerol solution layer by layer (Cui et al 2022).

Reactive extrusion- In this process polymerization reaction is carried out with minimum solvent with a rapid reaction of 1 to 15 minutes such as cross-linking, glycolysis, hydrolysis, etc. Twin screw extruders are used as chemical reactors (Garcia-Guzman et al., et al. 2022).

Bioplastics

Plastics have been used widely in packaging material, electrical and electronic devices, industrial machinery, etc. These plastics are generally synthetic and are produced by polymerizing petroleum derivatives. They are diverse and very versatile in their applications, indicating that more than 700 types of plastics are currently produced, such as polystyrene, nylon, polyurethane, polyvinyl chloride (PVC), silicones, epoxy resins and polyamides. However, this versatility is accompanied by several ecological disadvantages: inability to biodegrade, high resistance to corrosion and water, and greenhouse gas emissions.

Despite this frightening scenario, one of the alternatives gaining momentum is the replacement of plastics with bioplastics, as they do not form a single class of polymers, but rather a group of materials with different properties and a range of applications.

An investigation conducted by Jason Chen on June 2014 for BCC Research (Chen 2014) predicted an annual growth rate of 32.7% for bioplastics from 2014 to 2019. In this study, Asia and South America are expected to account for about 90% of bioplastics production capacity in 2019 compared to 65% in 2013. Sugarcane fiber present in residues such as bagasse, leaves and buds has great potential for development in the field of bioplastics. The degree of commercialization of bioplastics based on renewable resources such as fibers has reached different levels. Some of them have already reached a certain level of maturity that allows them to be competitive with petroleum-based plastics. The price differences that isolated them from conventional resins have been reduced significantly. In addition, factors such as high oil costs, increasing advances in the processing of these types of materials, a growing number of consumers concerned about environmental protection, and the creation of new government legislation have allowed the consolidation of specific applications of bioplastics at the commercial level. In addition, competition has increased in recent years, especially for polymers derived from cellulose and starch except for polylactic acid (PLA), which will show a slower growth rate of 26.6% and 18.1% for the first two and 16.7% for later. In contrast, bioplastics such as polyethylene (Bio-PE) and polyethylene terephthalate (PET) have seen significant growth in the last few years.

The classification of bioplastics according to their origin, as biopolymers or from petrochemical biodegradables, can be observed in Fig. 1.

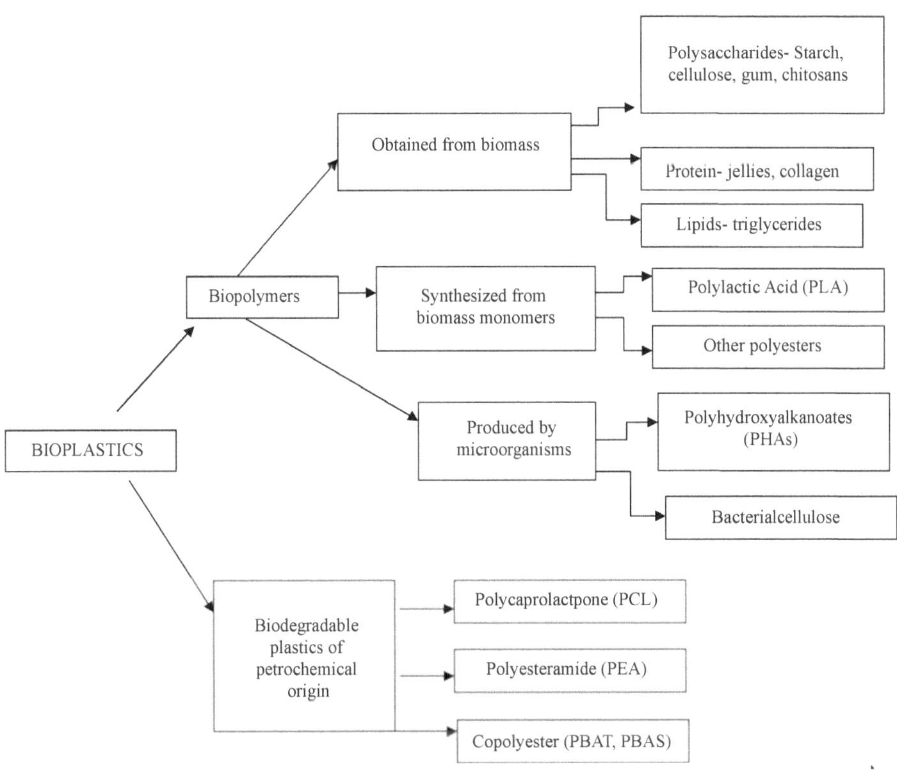

Fig. 1: Classification of bioplastics according to their origin

Starch-based bio foams

Foam plastic namely Polyurethane (PU), polystyrene (PS), and polyethylene (PE) are the main segment of foam plastic and cover a large market share. Many countries have imposed a ban on disposable plastics. Starch can be a prospective substitute for this category with some modification. Starch-based foaming agents can be formed when the water present in the open cell evaporates quickly leading to the formation of rigid foam (Jiang et al., 2020). These starch foams can replace traditional plastic foams (PS, PE, and PE). In foam preparation melting temperature is important usually water, xanthan gum, etc. are used as a plasticizer. In starch-based material, lower water content and higher temperature of starch results in open-cell structure leading to evaporation of water and facilitates foam making process. These foams can be in a ball shape to sheet shape or as per the requirement of packaging materials. Foaming materials are 90% gas hence compounded pellets are easier to transport (Soykeabkaew et al., 2015).

Biocomposite and Nanocomposite of Starch

To develop starch-based material into packaging material the greatest challenge is the reactivity towards moisture or its hydrophilic nature. To improve the mechanical properties of starch, it is either modified or blended with hydrophilic polymers. These blendings maybe with natural fibers, rubber, cellulose, or synthetic polymers like PP, PLA, and PHB (Noivoil and Yoksan, 2021, Garcia-Guzman et al., 2022). The process reduces via esterification of the -OH group of α-D glucose with methylene groups. Usually reinforcement of organic and inorganic is in small concentration into a starch matrix. Recently wheat hull, corn hull, kenaf, and walnut shell, the product of agriculture Industries have been successfully used as a composite in starch. Modern technology and processing techniques have manipulated matters at the nanoscale involving the synthesis of improved materials. Usually, natural fiber, glass fiber, paper fibers, cellulose nanofibers, graphite, silica, clays, etc. are used in nanocomposites having a surface area between 0.52-500 m^2/g. TPS (glycerol) matrix is used with cellulose and clay. Clays are colloidal fragments of silicates made up of tetrahedral and octahedral sheets of cations (Al, Mg) and co-ordinate by oxygen atoms (Dufresne et al., 2015). Recently nano-silica (SiO_2), bentonite, and hectorite clays have been successfully incorporated into TPS-based composite with good mechanical strength and improved biodegradability properties. Corn starch/clay films showed the highest absorption and good antibacterial properties (Teoh et al., 2012). Recently starch-based edible film with cellulose microcrystals had shown promising results with good tensile strength and increased thermal stability (Jiang et al., 2022).

The application of starch-based functional materials is being used for packaging extract of pomegranate peel, clove extract in improving the mechanical strength and antibacterial properties of the film and is known as functional composite.

The major challenge in starch-based biomolecules for packaging material is its water affinity and poor mechanical strength however very promising results have been seen with the technological intervention and further opportunity exists in decreasing cost and improvement in quality parameters.

Biochemical

Under this name, we consider biochemicals to be a set of chemicals produced from biomass. It is important that throughout the production of chemicals and indeed throughout the life cycle of the product, energy consumption must be minimized, the processes used must be safer, and the use of hazardous chemicals and their production must be avoided as preconceived principles of green chemistry (GC). The bio-based chemicals market is expected to grow due to business initiatives that favor the use of sustainable and renewable raw materials. These products are considered "building blocks" or chemical platforms, that is, they have the potential to produce more derivatives. These chemical products are produced through the biological and chemical conversion process. The chemicals like succinic, fumaric, and malic are produced through a biological process and utilized as fiber (Lycra), polymers anti-freezing substances, food additives, and flavoring substances. Aspartic acid, gluconic acid, glutamic acid, and itaconic acid are synthesized through biological conversion and have applications in textile (absorbent polymers, polyesters, amides, diapers, contact lenses), paints solvent, corrosion inhibitors, water treatment systems, adhesives, lubricants), agriculture and food (herbicides, flavors) industries. Glycerol produced through both chemical and biological processes had wide applications as a disinfectant, lubricants, cosmetic products, polyester, polyurethane resin, etc. Sorbitol, 2-5 Furandicarboxylic acid, and xylitol are produced through chemical methods and have applications as sweetener and PET, film, polyester, nylon, etc respectively. Xylitol is used in mouth wash, toothpaste, hard candies, etc. (Xu et al, 2019)

Several aspects including high oil prices, consumer preference for products from renewable resources, business commitments, mandates, and government support are driving efforts to broaden the spectrum of biochemicals. In the mid-to long term, biochemical products will share the market with petroleum-based chemicals and eventually replace them as competitively priced products after overcoming the technological challenges they face today.

According to research done by the U.S. Department of Agriculture (2017), the degree of research, development, and commercialization of chemicals based on renewable resources is at different levels. The most highly developed biochemicals are derivatives of 1,4-diacids (succinic, fumaric, malic), as well as xylitol, sorbitol, and y-3-HP. The world demand for these acids was approximately $832 million in 2013 (Evans 2011).

Pharmaceutical-based Bioproducts from Starch

Starch in earlier days had a stigma of having hyperglycemic properties. Research in the past few decades has shown that starch based on its digestibility can be broadly classified into three i.e., rapidly digestible starch (RDS), slowly digestible starch (SDS), and resistant starch (RS). Starch in the human body is acted upon by salivary and pancreatic alpha-amylase and is broken down to glucose and maltose. The starch which is not digested and reaches to large intestine is not utilized as energy in the body and resembles to fibers. Resistant starch (RS) is that fraction of starch that is resistant to hydrolyses by amylase (in vitro) after 120 minutes of incubation or escapes digestion in the small intestine and reaches to colon and can be fermented by gut microbiota (Dona et al., 2010). Raw potato, banana, pasta, etc., are good sources of resistant starch.

RS are further classified into five categories namely RS1, RS2, RS3, RS4, and RS5 which may show resistance due to their dense structural configuration or protected by cell wall inhibiting the exposure of embedded starch with amylase enzyme. RS1 and RS2 are slow rate starch and very slow RS respectively where starch is protected by a cell wall or has a crystalline structure. Retro graded or modified starches have a slow rate of digestion which resist hydrolyses and are classified as RS3 and RS4 respectively. RS5 are defined as those water-insoluble polysaccharides that are linear poly alpha- 1,4- D glucan not affected by alpha-amylases and have therapeutic properties or prebiotic properties as they are converted into short chain fatty acids (SCFA) in the colon (Aziz 2013).

Table 3: Resistant Starch and their sources

S.No.	Type of Resistant Starch	Source	Content	Description	References
1.	RS1	Coarsely ground or whole-kernel grains	<2.5%	Physically inaccessible starch	Birt et al. (2013)
2.	RS2	High-amylose maize starch, raw potato, raw banana starch	<2.5%	Granular starch with the B- or C-polymorph. This starch displays a high gelatinization temperature, above the boiling point of water. After boiling or cooking at a temperature below its gelatinization temperature, this type of starch retains its crystalline structure and remains resistant to enzymatic hydrolysis.	Birt et al. (2013)
3.	RS3	Cooked and cooled starchy foods	5.0–15%, The highest amount of total RS was obtained for the retrograded gelatinized starch.	Retrograded starch. After starchy foods are stored, particularly in a refrigerator, amylose molecules and long branch chains of amylopectin form double helices and lose their water- binding capacity.	Birt et al. (2013), Tacer-Caba, et al. (2018)

S.No.	Type of Resistant Starch	Source	Content	Description	References
				The double helices of starch molcculcs do not fit into the enzymatic binding site of amylase, thus they cannot be hydrolyzed by this enzyme.	
4.	RS4	Cross-linked starch and octenyl succinate starch. Some fibre-drinks, foods in which modified starches have been used (e.g. certain breads and cakes).		Chemically modified starches. Highly cross-linked starch remains in a granular form after cooking, with little enzymatic susceptibility, and cannot be hydrolyzed by amylases or fermented by microbes.	Birt et al. (2013), Jyothsna et al. (2017)
5.	RS5	Stearic acid-complcxcd high-amylose starch		Amylose-lipid complex. Thc amylosc-lipid complex also entangles amylopectin molecules, restricting the swelling of starch granules and enzyme hydrolysis (30,31). Because the amylose-lipid complex formation is an instant reaction and the complex can reform after cooking, type V resistant starch (RSV) is considered thermally stable.	Birt et al. (2013)

Starch in its native or modified form has very wide application in the pharmaceutical industry. As binder and diluent, starch helps manufacture viscous paste in the wet granulation process to improve the flow of the drug. This prevents powder segregation and irregular flow. Secondly, drugs that are to be delivered in small doses require inert material. Starch has been bland in taste and can be a very good bulk material that can be eventually compressed into tablets along with required doses.

Starch plays a good role as disintegrant and absorbents. This is an important step in drug release and absorption. Dissolution or disintegration into small particles helps in the swelling of particles in presence of water. Starch easily gets disintegrated due to its weak associative force and formation of channels allowing liquid to penetrate solid matrix. In addition, starch is a very good absorbent due to its hygroscopic nature, drugs that are liable to hydrolyses can retain stability by adding starch (Olobayo et al., 2019)

Starch can also be used as microencapsulation material. Encapsulation is a technique in which any desired particle is enclosed within a matrix to conserve its degradation and release. The particle may be an active ingredient, a probiotic, or vitamins, lipid, protein, etc., and are usually used as coating materials. Starch particularly RS2 has been successfully used as encapsulating material for probiotics (Sultana, 2000). Microencapsulation of probiotics with starch and calcium alginate mixed at 2% rate in ice-cream samples stored at -20°C had shown 30% improvement in result (Aziz, 2013).

Health Benefits

A high intake of carbohydrates is associated with hyperglycaemic properties; however, RS has half the calories (8KJ/g) in comparison to digestible starch (Aziz, 2013). Physiologically resistant starch has almost all the beneficial effects of dietary fiber. Recently, RS has been well studied for its positive role as a prebiotic. Probiotics can ferment RS and produces short-chain fatty acids (SCFA) which have a beneficial role in lowering cholesterol and preventing pathogenic bacteria etc, (Aziz, 2013). Studies have reported that RS help in stimulating gut bacteria such as Lactobacilli, and Bifidobacteria (Wanikar and Kotwal, 2021). It has also been reported that RS when gets fermented by colonic microbiota reduces more amount of butyric acid in its SCFA content. Butyric acid has been associated with less formation of cancer-producing colonic cells. Thus, RS also helps in the prevention of colon cancer (Wanikar and Kotwal, 2021).

Application of Starch in Food Products

Starch is a highly flexible polymer, and there are several ways to modify its structure and obtain a functional product with desired properties for specific industrial applications and increase its added value. The reaction of starches is exploited to create different types of modified/converted starches with appropriate physiochemical characteristics such as gelatinization retrogradation, heat stability, solubility, color, texture, and so on so forth. The industry of starch modification is constantly evolving. The food industry is very mindful of the safety of chemical residues hence not all types of modified starches are used in foods. Generally, modified starches are used for adhesion and as a binder in breaded and battered foods, used as meat and snack seasonings, as well as dustings for chewing gum and products produced in the bakery, moreover as crisping cover for fried snacks and salad dressings, emulsion stabilizers, creamers, foam stabilizer, canned foods and as expanders in baked snacks and cereal meals (Mason, 2009). Chemically modified starches have wide applications in baked products, soups, RTE products, ice cream, beverages, biscuits, etc.

Baked Products (bread, pies, wafers, biscuits)

Baked foods containing modified starch as a functional ingredient include cake formulas, glazes, and fillings. Pregelatinized starch is used in cake mixes to soften the cake and retain moisture in the baked product. They are also used to provide convenience food (Ashogbon et al. 2014). Pregelatinized starches are also used in cookies to control spread. Film-forming starches are added to glaze mixes to provide an attractive coating, moisture barrier and decrease tackiness. Starches used include dextrins, stabilized hydrolysates, and maltodextrins. They also provide resistance to boil-out during baking and resistance to weeping during distribution (Park et al., 2018). Pregelatinized derivatized starches are sometimes used to avoid the costs of heating and cooling fillings in bakery products (Majzoobi et al., 2011).

Gravies, Soups, and Sauces (soups, sauces, tomato paste, or ketchup)

Esterified and etherified starches are used for making gravies, soups, and sauces. Esterification, particularly with acetylation of starch with an acetic acid anhydride, produces starch acetate which can be used as a bulking agent in snack foods, stabilizer, and thickener in most foods improving the smoothness and sheen of soups and sauces or ketchup, cholesterol-free salad dressings, and flavor encapsulating agents, clouding agents, creamer, and beverage. It can be used as a substitute for gum Arabic, egg yolk, and caseinates. Esterification of starch reduces rancidity by preventing oxidation and increasing viscosity. Similarly, etherified starches have improved clarity of starch paste, greater viscosity, reduced syneresis, and freeze-thaw stability (Egharevba, 2019).

Salad dressing, Mayonnaise, Ice Cream, Spreads, and Beverages

Hot-process salad dressings are typically made by cooking starch, vinegar, salt, sugar, and water to a paste and then cooling the paste. The oil is added last, typically with egg yolks, whites, or other emulsifiers. A more stable emulsion is achieved by sending the complete formula through a colloid mill, a very high shear process (Saboonchia et al., 2021).

Cold-process dressings follow a similar flow, but without heating. Pourable dressings are more often made via cold process. Modified starch is used in dressings to thicken and stabilize the dressing and to provide the desired flow character. Because of the low pH, high temperature, and high shear involved applied in cook-up dressings requires the starch to be highly crosslinked (Saboonchia et al., 2021). Modified waxy maize starch products alone are more typical in pourable dressings. Potato and tapioca-modified starches have not been successful in this application, due to their greater susceptibility to acid and shear. Fat-reduced and fat-replaced dressings generally use a more highly stabilized, crosslinked waxy maize starch (Maani et al., 2017).

Succinylation of starch with OSA (octenyl succinic anhydride) produces beverage emulsion stabilizers and mayonnaises, flavor encapsulating agent for battered meat and meat products (Manson, 2009). Etherification treatment to starch especially hydroxypropylation produces hydroxy populated starch that can be used in ice creams, refrigerated and frozen foods, and dairy products (Manson, 2009). Crosslinked starches can be used in dips, spreads, fruit fillings, and puddings (Korma et al., 2016).

A beverage emulsion is a concentrate with added sugar and carbonated water. The oil-in-water emulsion provides flavor as well as opacity in products such as orange soda. Modified starch has been used to stabilize these emulsions and prevent creaming (phase separation), sedimentation, and loss of flavor. The concentrate is made by homogenizing the oils with an equal amount of the solubilized starch, citric acid, sodium benzoate, and color where desired. A fine emulsion required a kind of modified starch for better physical and mouthfeel properties (Greer, 2018).

Confectionery (Candy, Sweets, and Sweetmeat)

Oxidized starches are more suitable for confectionery items due to their low-temperature stability, high clarity or transmittance, and low viscosity. It can be used for coating candies and sweets since they easily melt (Egharevba, 2019).

Puddings (Custard, Pap, others)

For the preparation of puddings and custards, etc. pregelatinized starches have been used to achieve thickening and water retention without applying heat. It has also been used in ready to use bread mixtures. Increased retention and absorption of water enhance the quality of products as an agglutinant in the meat industry and as a filling for fruit pies (Egharevba, 2019).

Innovative Products

Modified starch is used in retorted baby foods to thicken and provide a short texture. In addition to high-temperature stability, fruit-based, dessert-type products require resistance to low pH and stability (Vanier et al., 2020). Accordingly, crosslinked acetylated waxy maize and tapioca starches are favored in this sensitive product. Tapioca starch products are favored for flavor because they are thought to be more healthful for babies (Woo and Seib, 2002).

Modified Starch is most commonly used in to thicken, stabilize and enhance the mouthfeel of canned foods. It is also used to increase the viscosity at the time of filling to lessen suspended particulates and control splashing (Masina et al., 2017). The cross-linked starches are used to thicken many canned foods like waxy maize starch. Canning starches will contain various levels of monosubstitution depending on the stability and food texture (Woo and Seib, 2002). Fill-viscosity modified starches with or without monosubstitution or acid thinning, both enhance breakdown. These give high initial viscosity for suspension of particulates during the filling process. (Masina et al.2017).

Modified starch is used in frozen foods for the same reason as it is used in refrigerated or canned foods, generally for thickening, temperature stability, and control of the flow in the purpose food. Freezing intensifies syneresis in the food system Freezing introduces stability demands beyond the other forms of distribution (Chaudhuri et al., 2020). The stability of the starch is critical to the stability of frozen food. Modified starch is used in frozen battered foods to provide better adhesion and decreased oil pick-up during frying, and also for improved crispness after oven or microwave reconstitution (Wang and Xu et al., 2018). Oxidized starches are used for improved adhesion reputedly because of covalent bonding with protein on the substrate surface. High-amylose starches and flours are favoured for crisping.

Conclusion

Starch is abundantly present in plant sources. Due to its chemical composition, it can be easily degraded by microorganisms hence its utilization for the development of the various industrial products will not impose any harmful effect on the environment. This process is also utilized in the production of bioenergy in the form of bioethanol and other fuels like biohydrogen, methane, etc. Starch and its structure are dependent on the sources from where it is obtained. Its physio-chemical properties can be manipulated by various techniques to obtain various useful bioproducts of industrial importance. In the evolution of its utilization, thermoplastic starch (TPS) are the initial bioproduct utilized in various packaging materials and blends of starch compounds are developed using natural and synthetic polymers. In recent years, starch-based nanocomposites and nanostructures have shown great potential in the development of active films, gels, stabilizing emulsions as well as encapsulating and binding agents.

References

Ashogbon, A. O., &Akintayo, E. T. (2014). A recent trend in the physical and chemical modification of starches from different botanical sources: A review. Starch-Stärke, 66, 41-57.

Babu, S & Parimalavalli, R. (2014). Effect of starch isolation method on properties of sweet potato starch. The Annals of the University Dunarea de Jos of Galati Fascicle VI – Food Technology, 38(1), 48-63.

Balat, M., & Balat, H. (2009). Recent trends in global production and utilization of bio-ethanol fuel. Applied energy, 86(11), 2273-2282.

Balat, M., Balat, H., & Öz, C. (2008). Progress in bioethanol processing. Progress in energy and combustion science, 34(5), 551-573.

Bao J, Ao Z, Jane J. (2005) Characterization of physical properties of flour and starch obtained from gamma-irradiated white rice. Starch. 57:480-487.

BeMiller JN, Huber KC. (2015) Physical modification of food starch functionalities. Annual Review of Food Science and Technology. 2015; 6:19-69.

Bertolini AC, Mestres C, Colonna P, Raffi J. (2001)m Free radical formation in UVand gamma-irradiated cassava starch. Carbohydrate Polymers. 44: 269-271.

Betancur AD, Chel GL, Cañizares HE. (1997) Acetylation and characterization of Canavalia ensiformis starch. Journal of Agricultural and Food Chemistry. 45:378-382.

BiELACZYC, P., Woodburn, J., Gandyk, M., & SZCZOTkA, A. (2016). Ethanol as an automotive fuel–a review. Combustion Engines, 55.

Birt, D. F., Boylston, T., Hendrich, S., Jane, J. L., Hollis, J., Li, L,, & Whitley, E. M. (2013). Resistant starch: promise for improving human health. Advances in nutrition, 4(6), 587-601.

Boukhalfa F, Kadri N, Bouchemel S, Ait Cheikh S, Chebout I, Madani K, et al. Chemically (2018) Modified starch and utilization in foodstuffs. Mediterranean Journal of Nutrition and Metabolism. 11:37-50.

Cappelletti BM, Reginatto V, Amante ER, Antônio RV. (2011) Fermentative production of hydrogen from cassava processing wastewater by *Clostridium acetobutylicum*. Renewable Energy. 36:3367–3372.

Cardona, C. A., & Sánchez, Ó. J. (2007). Fuel ethanol production: process design trends and integration opportunities. Bioresource Technology, 98(12), 2415-2457.

Chaudhuri, S. D., Mandal, A., Dey, A., & Chakrabarty, D. (2020). Tuning the swelling and rheological attributes of bentonite clay modified starch grafted polyacrylic acid-based hydrogel. Applied Clay Science, 185, 105405.

Chen, Y. J. (2014). Bioplastics and their role in achieving global sustainability. Journal of Chemical and Pharmaceutical Research, 6(1), 226-231.

Čížová A, Sroková I, Sasinková V, Malovíková A, Ebringerová A. (2008) Carboxymethyl starch octenylsuccinate: Microwave-and ultrasound-assisted 18 Chemical Properties of Starch synthesis and properties. Starch. 60:389-397.

Cui, Y., Li, C., Guo, Y., Liu, X., Zhu, F., Liu, Z., Liu, X., & Yang, F. (2022). Rheological & 3D printing properties of potato starch composite gels. Journal of Food Engineering, 313, 110756. https://doi.org/10.1016/j.jfoodeng.2021.110756

da Rosa ZE, Dias ARG. (2011) Impact of heat-moisture treatment and annealing in starches: A review. Carbohydrate Polymers. 83:317-328

Das Neves, M. A., Kimura, T., Shimizu, N., & Nakajima, M. (2007). State of the art and future trends of bioethanol production. Dyn Biochem Process Biotechnol Mol Biol, 1(1), 1-14.

Demirbas, A. (2016). Conversion of oil shale to liquid hydrocarbons. Energy Sources, Part A: Recovery, Utilization, and Environmental Effects, 38(18), 2698-2703.

Dona AC, Pages G, Gilbert RG, Kuchel PW (2010). Digestion of starch: In vivo and in vitro kinetic models used to characterise oligosaccharide or glucose release. Carbohydrate Polymers.80:599-617.

Dufresne, A. (2014). Crystalline starch-based nanoparticles. Current Opinion in Colloid &Interface Science, 19, 397–408. https://doi.org/10.1016/j.cocis.2014.06.001

Egharevba H.O. (2019). Chemical Properties of Starch and Its Application in the Food Industry. DOI: 10.5772/intechopen.87777.

Eliasson A-C, Kim HR. (2008) Changes in rheological properties of hydroxypropyl potato starch pastes during freeze-thaw treatments I. A rheological approach for synthesis and properties. Starch. 60:389-397.

Evans J (2011) Market research. Available on: https://www.bccresearch.com/market-research/biotechnology/commercial-amino-acids-bio007j.html. accessed on September 2022.

Fang HHP, Li C, Zhang T. (2006) Acidophilic biohydrogenproduction from rice slurry. International Journal ofHydrogen Energy.31:683–692.

Fiedorowicz M, Tomasik P, You S, Lim S-T. (1999) Molecular distribution and pasting properties of UV-irradiated corn starches. Starch. 51:126-131.

García-Guzmán, L.; Cabrera-Barjas, G.; Soria-Hernández, C.G.; Castaño, J.; Guadarrama-Lezama, A.Y.; Rodríguez Llamazares, S. (2022) Progress in Starch-Based Materials for Food Packaging Applications. Polysaccharides 3:136-177. https://doi.org/10.3390/polysaccharides3010007

Gifuni, I.; Olivieri, G.; Krauss, I.R.; D'Errico, G.; Pollio, A.; Marzocchella, A. (2017) Microalgae as new sources of starch: Isolation and characterization of microalgal starch granules. Chem. Eng. Trans. 57, 1423–1428.

Global Industrial Starch Market Size Report, https://www.grandviewresearch.com/industry-analysis/industrial-starch-market-report accessed on September 2022.

Greenwood CT, Muir DD, Whitcher HW. Hydroxyethyl starch as a cryoprotective agent for human red blood cells the relation between the molecular properties and the cryoprotective effect. Starch. 1977;29: 343-347

Greer, F. R. (2018). Use of starch and modified starches in infant feeding: a historical perspective. Journal of Pediatric Gastroenterology and Nutrition, 66, S30-S34.

Hansen, M. R., Blennow, A., Pedersen, S., Nørgaard, L., &Engelsen, S. B. (2008) Gel texture and chain structure of amylomaltase-modified starches compared to gelatine. Food Hydrocolloids 22, 1551-1566.

Hernández AR. (2018) Chemical modification of starch with synthetic. Applications of Modified Starches. 2:3-22.

Homayouni, A., Amini, A., Keshtiban, A. K., Mortazavian, A. M., Esazadeh, K., & Pourmoradian, S. (2014). Resistant starch in food industry: A changing outlook for consumer and producer. Starch-Stärke, 66(1-2), 102-114.

Huang Z-Q, Lu J-P, Li X-H, Tong ZF. (2007) Effect of mechanical activation on Physico-chemical properties and structure of cassava starch. Carbohydrate Polymers. 68:128-135

Hussy I, Hawkes FR, Dinsdale R, Hawkes DL. (2003) Con-tinuous fermentative hydrogen production from a wheat starch coproduct by mixed microflora. Biotechnology and Bioengineering 84:619–626

Jiang, C., Rashid, M.A.R., Zhang, Y. et al. (2022) Genome-wide association study on development and evolution of glutinous rice. BMC Genom Data 23, 33. https://doi.org/10.1186/s12863-022-01033-1

Jiang, Q. Duan, J. Zhu, H. Liu, L. Yu, (2020) Starch-based biodegradable materials: challenges and opportunities, Advanced Industrial and Engineering Polymer Research 3, 1: 8–18.

Jyothi AN, Moorthy SN, Rajasekharan KN. (2006) Effect of cross-linking with epichlorohydrin on the properties of cassava (Manihot esculenta Crantz) starch. Starch. 58:292-299.

Jyothsna, E., & Hymavathi, T. V. (2017). Resistant starch: Importance, categories, food sources and physiological effects. J. Pharmacogn. Phytochem, 6(2), 67-69.

Kaper, T., van der Maarel, M. J. E. C., Euverink, G. J. W., & Dijkhuizen, L. (2003) Exploring and exploiting starch modifying amylomaltases from thermophiles. Biochemical Society Transactions 32, 279-282.

Karim, A., Tie, A.P.-L., Manan, D. and Zaidul, I. (2008), Starch from the Sago (Metroxylonsagu) Palm Tree—Properties, Prospects, and Challenges as a New Industrial Source for Food and Other Uses. Comprehensive Reviews in Food Science and Food Safety, 7: 215-228. https://doi.org/10.1111/j.1541-4337.2008.00042.x

Katopo H, Song Y, Jane J. (2002) Effect and mechanism of ultrahigh hydrostatic pressure on the structure and properties of starches. Carbohydrate Polymers. 47:233-244.

Kong, I. (2012). Thermoplastic Starch. (2015) Thermoplastic Elastomers.Soykeabkaew, N.; Thanomsilp, C.; Suwantong, O. A review: Starch-based composite foams. Compos. Part A Appl. Sci. Manuf. 78, 246–263.

Korma SA, Kamal-Alahmad NS, Ammar A-F, Zaaboul F, Zhang T. (2016) Chemically modified starch and utilization in foodstuffs. International Journal of Food Sciences and Nutrition. 5:264-272.

Krithika, L. &Ratnamala, K.V. (2019). Modification of starch: a review of various techniques P. IJRAR- International Journal of Research and Analytical Reviews. Volume 6 I Issue 1pp 32-45.

Kulp K, Lorenz K. (1981) Heat-moisture treatment of starches. I. Physicochemical Properties. 58: 46-48.

Kunle OO (2019). Starch source and its impact on pharmaceutical applications. Chemical Properties of Starch.29:35.

Lawal MV. (2019) Modified starches as direct compression excipients–effect of physical and chemical modifications on tablet properties: A review. Starch. 71:1800040

Liu H, Ramsden L, Corke H. (1999) Physical properties and enzymatic digestibility of hydroxypropylated ae, wx, and normal maize starch. Carbohydrate Polymers. 40: 175-182.

Liu J, Yang R, Yang F. (2015) Effect of the starch source on the performance of cationic starches having a similar degree of substitution for papermaking using deinked pulp. BioResources. 10: 922-931.

Loisel C, Maache-Rezzoug Z, Esneault C, Doublier J-L. (2006) Effect of hydrothermal treatment on the physical and rheological properties of maize starches. Journal of Food Engineering. 73:45-54

Maani, B., Alimi, M., Shokoohi, S., &Fazeli, F. (2017). Substitution of modified starch with hydrogen peroxide-modified rice bran in salad dressing formulation: physicochemical, texture, rheological, and sensory properties. Journal of Texture Studies, 48, 205-214.

Majzoobi, M., Radi, M., Farahnaky, A., Jamalian, J., Tongdang, T., &Mesbahi, G. (2011). Physicochemical properties of pre-gelatinized wheat starch produced by a twin drum drier. Journal of Agricultural Science and Technology, 13, 193-202.

Masina, N., Choonara, Y. E., Kumar, P., du Toit, L. C., Govender, M., Indermun, S., & Pillay, V. (2017). A review of the chemical modification techniques of starch. Carbohydrate polymers, 157, 1226-1236.

Mason WR. (2009) Starch use in foods. In: Starch: Chemistry and Technology. 3rd ed. (BeMiller JN editor). San Diego, California, USA: Elsevier Inc. Academic Press; 745-795.

Merlin A, Fouassier J-P. Etude de radicauxlibresforméas par irradiation ultraviolette de l'amidon: Application aux réctions de photodégradation et de photogreffage. Die MakromolekulareChemie. 1981; 182:3053-3068.

Michel JP, Raffi J, Saint-Lebe L. Experimental study of the radiodepolymerization of starch. Starch. 1980; 32:295-298.

Mittal, V. (2014), Functional Polymer Nanocomposites with Graphene: A Review. Macromol. Mater. Eng., 299: 906-931. https://doi.org/10.1002/mame.201300394

Mojović, L., Nikolić, S., Rakin, M., & Vukasinović, M. (2006). Production of bioethanol from corn meal hydrolyzates. Fuel, 85(12-13), 1750-1755.

Nabeshima EH, Grossmann MVE. Functional properties of pregelatinized and cross-linked cassava starch obtained by extrusion with sodium trimetaphosphate. Carbohydrate Polymers. 2001; 45:347-353.

Nawaz et al., 2013. Physical and Chemical Modifications in Starch Structure and Reactivity DOI: http://dx.doi.org/10.5772/intechopen.88870.

Nawaz H, Shad MA, Saleem S, Khan MUA, Nishan U, Rasheed T, et al. Characteristics of starch isolated from microwave heat treated lotus (*Nelumbo nucifera*) seed flour. International Journal of Biological Macromolecules. 2018;113:219-226

Nogueira, T., Cordeiro, D. D. S., Muñoz, R. A. A., Fornaro, A., Miguel, A. H., & Andrade, M. D. F. (2015). Bioethanol and biodiesel as vehicular fuels in Brazil assessment of atmospheric impacts from the long period of biofuels use. InTech, 1, 377-412.

Noivoil, N, Yoksan, R. Compatibility improvement of poly(lactic acid)/thermoplastic starch blown films using acetylated starch. J ApplPolym Sci. 2021; 138:e49675. https://doi.org/10.1002/app.49675

Olguin-Maciel, E., Larqué-Saavedra, A., Pérez-Brito, D., Barahona-Pérez, L., Alzate-Gaviria, L., Toledano-Thompson, T., Lappe-Oliveras, P., et al. (2017). Brosimum Alicastrum as a Novel Starch Source for Bioethanol Production. Energies, 10(10), 1574. MDPI AG. Retrieved from http://dx.doi.org/10.3390/en10101574

Omoregie E, Okugbo O. Various techniques for the modification of starch and the applications of its derivatives. Journal of Pharmacology Bioresource. 2015; 11:66.

Paleos CM, Sideratou Z, Tsiourvas D. Drug delivery systems based on hydroxyethyl starch. Bioconjugate Chemistry. 2017;28: 1611-1624.

Park, S. H., Na, Y., Kim, J., Dal Kang, S., & Park, K. H. (2018). Properties and applications of starch modifying enzymes for use in the baking industry. Food science and biotechnology, 27, 299-312.

Raffi J, Agnel JP, Saint-Lebe L, Dauberte B. (1981) Gamma radiolysis of starches derived from different foodstuffs. Starch. 33(Pt. 3): 269-271.

Saboonchia, S., Mehran, A., Bahramizadeh, P., &Massoud R. (2021). Application of modified starch in food. 3rd International Congress on Engineering, Technology, and Innovation. Germany.

Shaoxiao LWGZZ, Baodong Z. (2013) The influence of ultra-high pressure treatment on the physicochemical properties of areca taro starch. Journal of the Chinese Cereals and Oils Association. 2013; 5:80-84

Singh V, Ali SZ. (2000) Acid degradation of starch. The effect of acid and starch type. Carbohydrate Polymers. 41: 191-195.

Statista (2022) https://www.statistica.com/aboutus/our-research-commitment accessed on 22nd September 2022.

Sultana, K., Godward, G., Reynolds, N., Arumugaswamy, R., Peiris, P. and Kailasapathy, K., (2000). Encapsulation of probiotic bacteria with alginate–starch and evaluation of survival in simulated gastrointestinal conditions and in yoghurt. International journal of food microbiology, 62(1-2), pp.47-55.

Sweedman, M. C., Tizzotti, M. J., Schäfer, C., & Gilbert, R. G. (2013). Structure and physicochemical properties of octenyl succinic anhydride modified starches: a review. Carbohydrate Polymers, 92(1), 905-920.

Szymońska J, Wodnicka K. (2005) Effect of multiple freezing and thawing on the surface and functional properties of granular potato starch. Food Hydrocolloids. 19:753-760.

Tacer-Caba, Z., & Nilufer-Erdil, D. (2018). Resistant Starch. In G. Smithers (Ed.), Reference Module in Food Science Elsevier. https://doi.org/10.1016/B978-0-08-100596-5.22407-4

Tamaki S, Hisamatsu M, Teranishi K, Adachi T, Yamada T. (1998) Structural change of maize starch granules by ball-mill treatment. Starch. 50:342-348.

Tamaki S, Hisamatsu M, Teranishi K, Yamada T. (1997) Structural change of wheat starch granule by ballmill treatment. Journal of Applied Glycoscience. 44:505-513

Teoh, K.H., Ramesh, S. & Arof, A.K. (2002) Investigation on the effect of nanosilica towards corn starch–lithium perchlorate-based polymer electrolytes. J Solid State Electrochem 16, 3165–3170 https://doi.org/10.1007/s10008-012-1741-4

Tester RF, Debon SJ. (2000) Annealing of starch—a review. International Journal of Biological Macromolecules. 27: 1-12.

Thirumdas R, Trimukhe A, Deshmukh RR, (2017) Annapure US. Functional and rheological properties of cold plasma treated rice starch. Carbohydrate Polymers. 157: 1723-1731.

Usman, M.,TahirIshfaq, M., Malik, S.R., Iqbal, M and Ishfaq, B (2014). Alkaline Extraction of Starch from Broken Rice of Pakistan. International Journal of Innovation and Applied Studies ISSN 2028-9324 Vol. 7, No. 1, pp. 146-152.

Vanier, N. L., Pozzada dos Santos, J., Pinheiro Bruni, G., &Zavareze, E. D. R. (2020). Starches in Foods and Beverages. Handbook of Eating and Drinking: Interdisciplinary Perspectives, 897-913.

Vasantha, V. L., Sharvari, S., Alfia, N. S., & Praveen, N. (2021). Application of nanotechnology toward improved production of sustainable bioenergy. In Nanomaterials (pp. 445-479). Academic Press.

Vendruscolo, F. (2015). Starch: a potential substrate for biohydrogen production. International Journal of Energy Research, 39(3), 293-302.

Waliszewski KN, Aparicio MA, Bello LA, Monroy JA. (2003) Changes in banana starch by chemical and physical modification. Carbohydrate Polymers. 52:237-242

Wang, C., He, X., Fu, X., Huang, Q. & Zhang, B. (2016). “Substituent Distribution Changes The Pasting and Emulsion Properties of Octenylsuccinate Starch” in Carbohydrate Polymers, Vol. 135 pp. 64-71.

Wang, P., & Xu, X. (2018). Modified starches and the stability of frozen foods. Starch in food (pp. 581-593). Woodhead Publishing.

Wanikar, R. and Kotwal, S., (2022) Resistant Starch: A Promising Functional Food Ingredient. Starch: Evolution and Recent Advances, p.113. d.o.i 10.5772/intech.101558.

Wicaksono Y, Nuri N, Wisudyaningsih B. (2016) Effect of temperature and pH of modification process on the physical-mechanical properties of modified cassava starch. Molekul. 11:248

Woo, K. S., &Seib, P. A. (2002). Cross-linked resistant starch: Preparation and properties. Cereal Chemistry, 79, 819-825.

Xu, Y., Chi, P., Bilal, M., & Cheng, H. (2019). Biosynthetic strategies to produce xylitol: an economical venture. Applied Microbiology and Biotechnology, 103(13), 5143-5160.

Zanella Pinto V, Goncalves Deon V, Moomand K, Levien Vanier N, Pilatti Riccio D, da Rosa ZE, et al. (2019) Characteristics of modified carioca bean starch upon single and dual annealing, heat-moisture-treatment, and sonication. Starch. 71:1800173.

Zia-Ud-Din, Xiong H, Fei P. (2017) Physical and chemical modification of starches: A review. Critical Reviews in Food Science and Nutrition. 57:2691-2705.

4

Advances in Food Pigment Production

Poonam Yadav[1] and Amrita Poonia[2*]

[1]Department of Food Science and Technology, College of Agriculture G.B. Pant University of Agriculture and Technology, Pantnagar
[2]Department of Dairy and Science and Food Technology, Institute of Agricultural Sciences, Banaras Hindu University, Varanasi-221005 (U.P.) India

Abstract

The natural color of foods is due primarily to carotenoids, anthocyanins, betanin and chlorophylls, either as inherent food constituents or as food or feed additives. These compounds have drawn considerable attention in recent years, not because of their coloring properties, but due to their potential health-promoting effects. Their occurrence and levels in foods, along with the factors that influence the composition, have been widely investigated. Processing effects have been actively studied. In spite of the intense search for plant and microbial sources and efforts to increase yield, few natural food color additives have reached the market. Developing new colors for the food industry is challenging, as colorants need to be compatible with a food flavors, safety, and nutritional value, and which ultimately have a minimal impact on the price of the product. In addition, food colorants should preferably be natural rather than synthetic compounds. Micro-organisms already produce industrially useful natural colorants such as carotenoids and anthocyanins. Microbial food colorants can be produced at scale at relatively low costs. This chapter highlights the significance of color in the food industry, why there is a need to shift to natural food colors compared to synthetic ones and how using microbial pigments as food colorants, instead of colors from other natural sources, is a preferable option.

Introduction

Pigments are of a group of compounds that are intensely colored and are used to color other food materials. Color of the pigment is due to the chromospheres, a molecule specific structure which captures the sun energy and causes an excitation of electron from external orbital to higher orbital, where the non-absorbed energy is refracted or reflected to be captured by eye (Delgado-Vargas et al., 2000). In the present scenario, meaning of pigment is a substance constituted of small particles which is practically insoluble in the applied medium and is used due to its colorant, protective or other properties. It can also be defined as compounds which have uniqueness of characteristics to many companies. In the food industry, pigments used as food additives, antioxidants, color magnifier, etc. They reported in the broad range of colors, some of which are water-soluble and very less in number are insoluble in water. The terms pigment and color are mostly used for the food coloring materials, sometimes interchangeably (Babitha, 2009 and Malik, K. et al., 2012).

Until the mid-19th century all colorants were obtained from either plant or animal extracts. The textile industry used natural pigments, such as cochineal, wood madder, turmeric, or henna. In 1856, H. Perkin build up the first company which was responsible for the production of synthetic colors mauve. A few years later the other researched reported diazotization and a coupling reaction by Scientist Peter Griess was the next major step forward for development of the color industry. In the 19th century, synthetic organic dyes production was increased, creating a more economical and broader range of colorants. Since then, their quality has been enhanced due to extensive research and development (Sikorski, 2006 and Socaciu, 2007). The economic outcome of the color industry is clearly increased in the large number of amalgamate compounds; as much as 700 colorants are currently reported. They have conventionally been used in food as food additives, as a dye, cosmetic and pharmaceutical manufacturing companies; reported various hazardous problem (Durán, N. et al., 2002). All artificial foods suffered severe criticism, due to the presence of unnatural food additives and majorly different food color pigments. Today, all food color additives are very carefully regulated under the FSSAI (Food Safety and Standards Act, 2006) to make sure that foods are safe to eat and should properly labeled (Sikorski, 2006 and Socaciu, 2007). Pigments reported from natural sources are of worldwide interest and its gaining importance. These are preferred for their safe use as a natural food dye in substitute of artificial ones instead of having undesirable market (Sikorski, 2006). It is therefore, necessary to explore various essential natural sources of food grade pigments and their potential applications (Socaciu, 2007). Now a days. The use of natural pigment is reported as food additives, dye, as a cosmetic as well as in pharmaceutical companies (Venil, 2013 and Cristea, D and Vilarem, G. 2006). Natural food additives obtained from environmental ecosystem are found to be safer because of non-toxic, non-carcinogenic and biodegradable in nature. Some of the natural colors are also procured from native minerals, bug, herbs and pathogens (F. Delgado-Vargas and O. Paredes-López, 2002 and

Nagpal, 2011). Among all, microbial pigments are dominant sources. Pigment production from microbes as carotenoids, herbs or chemical synthesis, have problems of weather and geological differences in the production and as well in marketing Mata-Gómez, L.C. et al., 2014). Production of carotenoids from microorganism are of great interest because having the capability of stable pigment production and availability throughout year, they are of great interest owing to the stability of the pigments produced and the accessibility of refinement mechanization (L. Dufossé, 2006).

The great achievement in the production of pigment from microorganisms is that it provides easy and fast growth in a very cheap growth culture medium also independence from weather conditions and provides colors of various tincture. The economic advantages of microorganism pigment include growth on natural substrates such as saki rice wine, red bean curd and agro-industrial waste such as carbohydrate source (Mata-Gómez, L.C. et al., 2014). Microbial pigment has already been in used in fish industry, for example obtain good quality pink color salmon (Storebakken, T. et al., 1987). In nature, pigment producing microorganisms for example bacteria, fungus and yeast are very commonly used. Microorganisms has been reported for the production of various colorant like carotenoids, quinones, prodigiosins, melanins, flavins, and more specifically monascins, in literature (Vasanthabharathi, V., 2013 and Pantanella, F., 2007).

Carotenoids like β-carotene and xanthophylls such as astaxanthin are very important for healthy vision. It's also playing a crucial role in the suppression of different type of cancer and act as chemo-protective. In addition, sometimes β-carotene act as nutraceutical which alter carcinogenesis via anti-oxidation mechanism or by free radical generation (Venil, C and Lakshmanaperumalsamy, P., 2009).

The production of microbial pigments is very much affected by the temperature of incubation, depending upon the type of microorganism (Tinoi, J. et al., 2005 and Sanjay, K. et al., 2009). From growth and pigment production point of view, *Monascus* sp. produces pigment at 25-28 °C optimum temperature, whereas *Pseudomonas* species requires 35-36 °C optimum. The pH of the fermentation medium is another parameter that affect the growth and kind of pigment produced by the microorganisms in which they are grown (Joshi, V. et al., 2003). The yield of astaxanthin from *Phaffia rhodozyma* was 325 to 212 μg/g astaxanhin at a pH of 6.5 to 3.5 (Johnson and Lewis, 1979). Pigment production is also dependent on different type of carbon source either monosaccharides, disaccharides or polysaccharides etc. and different type of nitrogen source (Boonyapranai, K. et al., 2008). The optimization of growth conditions of microorganisms, particularly physical and nutritional parameters are of prime importance in the development of any pigment production process which will affect the economy and practicability of the process. Medium optimization and physical conditions have been customarily performed using one-factor-at-a-time method. The disadvantages of such a classical method are that it is time consuming, laborious and expensive; in addition, it fails to resolve the combined effect of different factors (Surwase and Aghaie-Khouzani, 2012). Statistical analysis is an important tool that can be utilized for the main data interpretation of interactive influences of different fermentation parameters on process performance. It is a useful way to collect information with limited experimental data, thereby limit the process performance time and cost (Myers, and Anderson-Cook, 2012). That's why researchers, now a day's are encouraged to apply statistical experimental methods such as Taguchi and response surface methodology (RSM), which provide a great amount of information based on only a small number of experiments (Ahmad, W. A. et al., 2012).

On summing up, the growing apprehension over the eventual side effects of artificial pigment on both the consumer health and the environment has raised preferential interest in natural coloring alternatives. Among all, microbial colorants popularly known as pigments have some advantages over plant and animal-based colorants. In lot of studies reported that microbes are known to produce a large number of stable pigments (Mussatto, et al., 2012). Large amounts of agro-industrial and domestic residues are generated from diverse economic activities; utilization of these residues as inexpensive substrates to support the growth of microorganisms to generate value-added products like pigments are of biotechnological interest in recent years (Joshi et al., 2003). Many research methods have been reported that a variety of cheaper substrates and wastes can be utilized as alternative substrates for the production of microbial pigments (Buzzini and Martini 2000). The utilization of several wastes as or raw materials notably helps in solving pollution problems, while their disposal may otherwise cause. As it is reported above, screening of different resources, establishing simple methodologies, and exploring the microbial synthesis of pigments on inexpensive substrates is an attractive option to develop commercial scale production (Delgado-Vargas and Paredes-López, 2002).

History of colorants

At the end of the 19th century, artificial colors were first used on a huge scale where they were considered as technological breakthrough. The term 'artificial' was linked with the idea of advance and artificial colorants were actually considered safer in food than the naturals, as they were tinctorially provides much powerful and consequently a small amount was required to obtain a specific-colored effect (Sikorski, 2006).

Artificial colorant has widely used in foods, pharmaceuticals, and cosmetics industry, but through the years their importance reduced. The reduction in usage due to the natural colorants started about five decades ago. All artificial colorant used in food as food additives received severe criticism. Artificial food colorant used as additives were one of the first man-made (synthetic) products regulated by law. Today, all food color additives are minutely regulated by Food Safety and Standards Act. 2006 to make sure that food contain limited quantity of artificial colorant which are not injurious to health (Freedman, 1978).

On the basis of their origin

Pigments can be categorized by their origin as natural or artificial. Natural colorants are generally produced by utilizing living organisms such as plants, animals, fungus or microorganisms. Artificial colorants are synthesized in laboratories by using chemical substances.

i. **On the Basis of Chemical Structure of the Chromophore Molecule:** These pigments can be categorized with the help chromophore chemical structure as chromophores are generally conjugated with carotenoids, anthocyanins, betalains, caramel and artificial pigments. Metal-associated compounds porphyrins, myoglobin, chlorophyll, and their derivatives are considered.

ii. **On the Basis of Structural Characteristics of the Natural Colorants:** Natural colorants can be classified by their structural characteristics as:

- Tetrapyrrole derivatives: It includes chlorophylls and heme colors.
- Isoprenoid derivatives: It consist of carotenoids and iridoids.
- N-heterocyclic compounds different from tetrapyrroles derivatives: purines, pterins, flavins, phenazines,
- Phenoxazines, and betalains.
- Benzopyran derivatives (oxygenated heterocyclic compounds) includes anthocyanins and other flavonoid pigments.
- Quinones: It consist of benzoquinone, naphthoquinone and anthraquinone.
- Melanins.

iii. **Pigment used as a food additive:** By considering the pigments as food additives, they are divided into two categories by Food and Drug Administration

Comes under certifiable categories: These are manmade and subdivided as artificial pigments and lagoon.

Comes under exemption from certification categories: This group includes pigments derived from natural sources such as herbs, ores, animals and manmade equivalent of natural derivatives.

Biological pigments-different classes

i. Biological pigments

In recent times, more concern of people toward utilization of natural edible coloring agents has decreased various synthetic coloring agents which have a potential of carcinogenicity and teratogenicity (Eyal and Spencer, 1992). Pigment from living organism is also well-known as simply pigments or biochromes. These are the constituents which are produced by living organisms which provide particular color absorption mechanism. The topic of synthetic dyes in food is in conversation for several years. The study and negative valuation of synthetic food dyes by the modern consumer have raised a strong interest in natural coloring substitute. Nature is rich in colors (minerals, plants, microalgae, etc.), and pigment-producing microorganisms (fungi, yeast, bacteria) are reasonably common (Babitha, S. et al. 2006 and Sheehan, H. et al., 2012). Till date all biological pigments reported, they absorb particular wavelengths of light while also reflect wavelength of light of other colors. The light which is absorbed by the living organism or plant are used to conduct chemical reactions, while the reflected wavelengths of light responsible for the color of the pigment that will appear to the eye. Biological pigments based on the source were categorized as mentioned below.

ii. Plant pigments

Several companies decided to use colors as food additives which are originated from plant extracts or pigments from plants. Plant pigments have a different variety of diverse type of molecules like carotenoids, porphyrins, betalains and anthocyanins. In plants, these pigments will also assist in pollination by attracting the pollinators. Some major plant pigments are as follows (Venil, C.K. et al., 2013):

a) **Chlorophylls**: These are the principal pigments in plants. These porphyrin compounds absorb yellow and blue wavelengths of the light and reflect the green. All green plants which are present on land and green algae possess two forms of this pigment chlorophyll a and chlorophyll b, whereas red algae possess only chlorophyll a.

b) **Carotenoids** : These are the orangc, rcd or yellow tetra-terpenoids. They gather the wave lengths that are not generously absorbed by chlorophyll molecules, most familiar carotenoids are carotene, lutein, lycopene. Carotenoids well studied to act as natural antioxidants.

c) **Anthocyanins:** These are water soluble flavonoid pigments that look like red to blue according to the pH. The anthocyanin catches the light that has passes the leaf and reflects it back to the regions having chlorophyll in order to maximize the use of available light. Color range of pink/red to mauve/blue were obtained from elderberries, black grape skin, black carrots, red cabbage etc.

d) **Betalains:** There are water soluble pigments like anthocyanins but differ from anthocyanins as they are indole-derived substances derived by tyrosine. These classes are only investigated in caryophyllales and does not observe in plants which contain anthocyanins. They provide the deep red color of beets and are widely used as food-coloring agents commercially.

Although plant pigments are emerging alternatives, they suffer from several bottlenecks i.e., for example; abundant orange-yellow pigment like curcumin (from plant rhizome of *Curcuma longa*) has to be debittered to avoid its odour and sharp taste. And pigments like anthocyanins, chlorophyll, betanin are pH-dependent, oxygen sensitive, heat sensitive, and subject to photo-oxidation some limitation of must be (Mata-Gómez, 2014).

iii. Animal pigments

Our ancestors managed to obtain wide range of pigments from animal sources before chemical equivalents are manufactured the most rare and difficult to obtain became symbol of wealth and status for example the color purple is associated with wealth and royalty. The purple dye of the ancients is reported as among the old pigments known in 13th century. Murexes types of drilling snails have a mucus secreting organ called hypobranchial gland. Tyrian purple is eked out in small amount from the mucus of certain mollusks. Similarly, caramine made from cochnieal insects is much more concentrated than the traditional red dye obtained from madder root. Carminic acid and carmine of orange to red and pink to red comes from a female cochineal insect from Peru and Equador. The pigment is price-sensitive here. Dye from kermes and cochneal was of high demand throughout Europe as it is used to color the fabrics of royalty, nobility and church leaders. For several centuries it was the dye used in British red coats, hand woven rugs, as paint. Carmine derived from cochneal is used to color drinks, foods, meat, sausages, processed poultry products, bakery products, pie fillings, icings, jams, desserts, yogurt, cheese, ice creams and other dairy products. Cosmetic industry is the major consumer of insoluble carmine pigment, particularly for hair and skin products, lipsticks, face powders, rouge and blushes. Another major application is in pharmaceutical industry to color ointments and pills. Lac insect produces red dye similar to that of the caramine dye, the water-soluble red dye comes from the body of the insect is obtained by aqueous extraction and is processed into seedlac and shellac. They have several applications including varnishes, paints, printing inks, sealing wax, coat pills, sweets and chocolates. Shellac is used in making vinyl records and to color Indian military uniforms and found in oriental carpets (Venil, C.K. et al., 2013). Other pigments like Canthaxanthin of orangish pink color are obtained from salmon, shrimp and flamingos. While several animal pigments are price sensitive, many areas having the main disadvantages such as limited color range and availability (Mata-Gómez et al., 2014).

iv. Microbial pigments

There are a number of natural pigments reported but only a few are available in proper amount for industrial production of different food items. Production of colorant from microorganisms is cheaper over other sources because microorganisms can grow fast which provide high productivity of the desired product. Development of new colors for the food industry is challenging, food additives used as color is necessary to be compatible with a food composition, food flavour, safety, and nutritional point of view, and which ultimately affect the consumer health and price of the product. Colorants which are used in food should be natural rather than artificial compounds. Micro-organisms has already reported for production of industrially important natural colorants like carotenoids and anthocyanins. Microbial food colorants are generally produced at relatively low costs. (Nagpal, N. et al., 2011 and Dufossé, L. et al., 2005).

a) **Fungal pigments:** It is reported that many fungi produce pigments during their growth phase which are mainly used as permanent staining of textiles and plastics. Some fungal pigments have properties like anthraquinone derivatives, resembling the important class of vat dyes. That's why fungal compounds have potential for the direct production of textile dyes or dye intermediates replacing chemical synthesis. The production and evaluation of microbial pigments as textile colorants is currently being explored (Hobson and Wales, 1998).

b) **Bacterial pigments:** Some bacterial species are able to produce pigments which is observed when they grow into colonies. These bacterial pigments can be used as a support during find out the bacterial species when they grow on growth media. For example, certain bacteria produce water soluble pigments which expanded through the medium in which they grow. Others bacterial pigments that are soluble in fat (Soliev, A.B. et al., 2011). It is documented that the yellow pigment zeaxanthin which is from *Flavobacterium* species can be used as an additive in poultry feed to fortify the yellow color of the skin of birds or to accentuate the color of the yolk of the egg. This yellow pigment zeaxanthin from *Flavobacterium sp.* is widely used in cosmetic and as well as in food industry. Canthaxanthin from the photosynthetic bacterium *Bradyrhizobium* sp. has been used in fish feed for numerous years. *Halobacterium* is also one more source of canthaxanthin, astaxanthin from *Agrobacterium aurantiacum* are used as colorant (Dufossé, L., 2006).

c) **Yeast pigments:** Yeast synthesizes a variety of commercially important carotenoids, viz., carotene, torulene, torularhodin and astraxanthin. Genera *Rhodotorula*, *Sporobolomyces* and *Phaffia* are potential sources of pigments. Carotenoids from yeast are used in pharmaceutical, chemical, food services industries. They are also an important precursor for vitamin A synthesis. Apart from coloring, it also has antioxidant and possible tumor-inhibiting properties. Few yeasts are reported that they are able to produce valuable carotenoids in pure culture on low-cost agro-industrial waste substrates, thereby providing an alternative to chemical fusion. Asthaxanthin from *Xanthophyllomyces dendrohous* formerly known as *Phaffia rhodozyma* is a carotenoid pigment which is used as a food colorant and widely used in the animal feed to impart color to the animal skin as animals have no capacity to synthesis the carotenoid pigment. There is many research have been reported till date as scientific papers and patents which deals with asthaxanthin production by utilizing this yeast still the colorant production process has not been economically efficient till now. Most of the research in recent years is also focused on *Rhodotorula glutinis* which provideds carotenoid pigment. However, some papers reported other species which are able to produce colorant such as *R. gracilis*, *R. rubra* (now *R. mucilaginosa*) and *R. graminis* too. The main compounds obtained by these red yeasts are torulene and torularhodin with minute quantity of beta carotene (Dufossé, L. et al., 2005).

Among reported pigment sources from microorganism have huge potential to produce diverse group of bio-products and one such bio-product is colorant materials. Now a days, interest in microbial pigments has increased considerably are mainly due to the benefits for human being and also due to the development of agriculture sector and veterinary industry (Nelis and Leenheer, 1991).

The production and their utilization as natural colorant from microorganism source have been documented by various many researchers. A large number of the microbial pigment productions are still at the R&D phase. And there are many studies in the literature on various microbial pigments which focus on production and their utilization of specified pigment in each case.

Efforts have been made in order to ease the production costs of microbial derived pigments compared to those of synthetic pigments or pigments extracted from natural sources. Development of new methodology will enhance the economy of pigment production by isolating totally new verities of microorganisms. Hence, work should be done on the microorganism pigments production by utilizing cheap, agro-industrial waste as a raw material source which will reduce the process cost and increase its application for cheaper industrial production (Nagpal, N. et al., 2011).

Classification of microbial pigments

Pigments produced by organisms as reminiscence of its secondary metabolism are commonly mentioned as bio-pigments. These bio-pigments have extensive synthetic and commercial application (Shirata, A. et al., 2011). Microbial pigments can be classified based on structural affinities and natural occurrence. Some examples of naturally occurring microbial pigments are:

i. **Riboflavin:** It is a yellow water-soluble vitamin B complex produced by various microorganisms. Traditional chemical synthesis of riboflavin is now being altered by commercially competitive biotechnological processes using another species ascomycetes *Ashbya gossypii*, filamentous fungi *Candida famata*, or bacterium *Bacillus subtilis*. It is also used in baby foods, breakfast cereals, sauces, pastas, processed cheese, fruit drinks, vitamin-enriched milk products, and some energy drinks (Stahmann, K.P. et al., 2000).

ii. **Beta-carotene:** *Phycomyces* and *Mucor circinelloides* (wild type) are reported a possible source of beta-carotene. The European Union Committee considers that beta-carotene obtained by fermentation of *Blakeslea trispora* is equivalent to the chemically synthesized material used as food additives colorant and is therefore acceptable for use as a pigment in foodstuffs (Dufossé, L. 2006).

iii. **Canthaxanthin:** It is produced by the major carotenoid pigment such as orange and dark pink-pigmented bacteriochlorophyll containing *Bradyrhizobium* (photosynthetic) strains which are isolated from stem nodules of *Aeschynomene species* and *Halobacterium sp.* Canthaxanthins and they are very effective antioxidants property and also inhibit thc of lipids oxidation (Dufossé, L. 2006).

iv. **Carotenoids:** They range between yellow to orange-red pigments which are ubiquitous in nature. A number of microorganisms has been reported in literature for production of carotenoids pigment like as *Serratia* and *Streptomyces*. They are strong antioxidants and widely used as food additives color. Few examples of microorganism which are investigated for the production of carotenoids belong to *Myxococcus*, *Streptomyces*, *Mycobacterium, Agrobacterium and Sulfolobus* (Eyal, and Spencer, 1992).

v. **Prodigiosin:** It is an all-round red pigment, and their production are reported by a number of microorganisms such as *Serratia marcescens*, *Vibrio psychoerythrus*, *Rugamonas rubra*, *actinomycetes*, such as *Streptoverticillium rubrireticuli* and other *eubacteria*. It also known to have property like antibacterial, anti-malarial, antineoplastic and antibiotic activity (Shirata, A. et al., 2000).

vi. **Phycocyanin:** It is a blue color pigment, which are produced by cyanobacteria containing chlorophyll a. The classification of blue colorant is also given on the name of spirulina which is a blue-green-algae. Cyanobacteria is also used as a dietary supplement which contain high amounts of proteins. Few supplements are reported which consists of dried cyanobacteria due to their high amount of protein (Malik, K. et al., 2012).

vii. **Violacein:** It is an all-round pigment obtained from bacterium *Chromobacterium violaceum* which exhibits numerous biological activities. It is popular due to importance in industrial markets, pharmaceuticals, cosmetic industry, food industry and textiles industry (Malik, K. et al., 2012).

viii. **Astaxanthin:** Chemically, it is known as 3, 3'-dihydroxy-β, β-carotene-4, 4'- dione. Astaxanthin is an orange-red pigment and their production are reported by microorganisms such as red *Basidiomycetous,* yeast *Xanthophyllomyces dendrorhous*, green algae *Heamatococcus pluvialis* and *Agrobacterium aurantiacum* (Rodríguez-Sáiz, M et al.,2008).

Pigment compounds form microorganisms based on color and appearance can be classified as shown in Table 1.

Table 1. Pigment producing microorganisms based on color and appearance

Microorganism (s)	Pigment/Molecule	Appearance	Reference
Bacteria			
Agrobacterium aurantiacum	Astaxanthin	Pink-red	Tinoi, J. et al., 2005
Paracoccus carotinifaciens	Astaxanthin	Pink-red	Storebakken, T. et al.,1987
Bradyrhizobium sp.	Canthaxanthin	Dark- red	Storebakken, T. et al.,1987
Flavobacterium sp., Paracoccus zeaxanthinifaciens	Zeaxanthin	Yellow	Sheehan, H. et al., 2012
Achromobacter sp.		Red	Dufossé, L. et al.,2005
Brevibacterium sp.	Carotenoid like	Orange yellow	Hobson and Wales, 1998
Corynebacterium michigannise		Greyish to cream	Soliev, A.B. et al., 2011
Corynebacterium insidiosum	Indigoidine	Blue	Soliev, A.B. et al., 2011
Rugamonas rubra, Streptoverticillium rubrireticuli, Vibrio gaogenes, Alteromonas rubra	Prodigiosin	Red	Johnson and Lewis, 1979
Xanthophyllomyces dendrorhous	Astaxanthin	Pink	Dufossé, L. 2006
Haloferax alexandrinus	Canthaxanthin	Dark Red	Nelis and Leenheer, 1991
Staphylococcus aureus	Staphyloxanthin Zeaxanthin	Golden yellow	Shirata, A. et al. 2000
Chromobacterium violaceum	Violacein	Purple	Stahmann, K.P. et al., 2005
Serratia marcescens, Serratia rubidaea,	Prodigiosin	Red	Asker and Ohta, 1999
Pseudomonas aeruginosa	Pyocyanin	Blue green	Malik, K. et al., 2012
Xanthomonas oryzae	Xanthomonadin	Yellow	Rodríguez-Sáiz, M et al.,2010
Janthinobacterium lividum	Violacein	Purple	Chen, Y. et al., 2003
Algae			
Dunaliella salina	β-carotene	Red	Couto and Sanromán, 2006
Chlorococcum sp.	Lutein	Orange	Sheehan, H. et al., 2012

Microorganism (s)	Pigment/Molecule	Appearance	Reference
Hematococcus sp.	Canthaxanthin	Yellow-orange-red	Venil, C.K. et al., 2013
Fungi			
Aspergillus sp.	β-carotene	Orange-red	Rodríguez-Sáiz, M et al.,2010
Blakeslea trispora	β-carotene	cream	Chen, Y. et al., 2003
Fusarium sporotrichioides	Lycopene	Red	Nagpal, N., et al., 2011
Haematococcus Pluvialis	Astaxanthin	Red	Myers and Anderson-Cook, 2009
Monascus sp.	Monascorubramin Rubropunctatin	Red-orange	Mata-Gómez, L.C. et al., 2014
Monascus purpureus	Monascin Ankaflavin	Red-yellow	Johnson and Lewis, 1979
Monascus roseus	Canthaxanthin	Orange-pink	Johnson and Lewis, 1979
Penicillium oxalicum	Anthraquinone	Red	Rodríguez-Sáiz, M et al.,2010
Blakeslea trispora	Lycopene	Red	Chen, Y. et al., 2003
Cordyceps unilateralis	Naphtoquinone	Deep-blood red	Sanjay, K. et al , 2007
Mucor circinelloides, Neurospora crassa and Phycomyces blakesleeanus	β-carotene	Yellow-orange	Sikorski, Z.E. 2006
Pacilomyces farinosus	Anthraquinone	Red	Dufossé, L. 2006
Yeast			
Cryptococus sp.	Melanin type compounds	Reddish brown	Nelis and Leenheer, 1991
Saccharomyces neoformans var. nigricans	Melanin	Black	Tinoi, J. et al., 2005
Phaffia rhodozyma	Astaxanthin	Pink-red	Shirata, A. et al. 2000
Rhodotorula sp. Rhodotorula glutinis	Torularhodin	Orange-red	Stahmann, K.P. et al., 2000
Yarrowia lipolytica	Melanin	Brown	Asker and Ohta 1999
Actinomycetes			
Streptoverticillium rubrireticuli	Prodigiosin	Red	Malik, K. et al., 2012
Streptomyces echinoruber	Rubrolone	Red	Rodríguez-Sáiz, M et al., 2010

Novel practices of microbial pigment production

Strain improvement

Improvement of the production strains provides the great opportunities for reduction of cost without significant mortgage in industries. Moreover, successful strain improvement depends by making and providing a fermentation industry more competitive ideas depends majorly on continuous improvement of the production strains. Major improvement in strains is generally done in increased yields of the desired metabolite production (Rodríguez-Sáiz, M et al., 2010). A more systematic and comprehensive approach for biotechnological process for strain improvement became more dominant now a days. Microorganism which are present in the environment are responsible for production of certain chemical compounds of biological interest. However, the industrial approach for the production these chemical compounds by environmental strains is not an economical process that's why wild strains are modified by the altering their genetic gene sequence or by regulating enzymes activity which resulted that specific product will be produced in excess amounts.

Improvement of microorganism genetic sequence by altering their gene sequence for the excess production of industrially important compound which benefited the commercial fermentation methodology. The strain improvement by using some common mutagens such as ultraviolet (UV), ethyl methane sulfonate (EMS) and 1-methyl-3-nitro-1-nitrosoguanidine (NTG) is convenient and can improve colorant yield as proved in several cases (Chen, Y. et al., 2003). Now-a-days, X-ray and fast neutron of irradiations are also used to obtain microorganism strain improvements. For example, *M. purpureus* was irradiated by using X-Ray to find out change in their pigmentation and morphological characterstics (Aghaie-Khouzani, M. et al., 2012).

Fermentation

Fermentation is a metabolic process that converts sugar to acids, gases and/or alcohol and usually occurs in yeast and bacteria. The production of microbial pigments by fermentation is an interesting area and a lot of attention is now paid to this biotechnological approach. This technique can be of solid or submerged fermentation mainly and each approach is having its own merits and demerits in pigment production.

i. **Submerged fermentation**

In Submerged fermentation, growth and decomposition of substrate (carbohydrate based such bagasse, molasses etc) occurs in the presence of excess amount of liquid broth media utilizes free flowing liquid substrates, such as molasses and broths. Different bioactive compounds are released into the fermentation broth media. The substrates are utilized rapidly by microorganism; hence there must be continuous feeding of growth nutrient required. Submerged fermentation methods are widely used for microorganisms which required high amount of moisture percentage hence high-water activity level. This type of fermentation methods makes the downstream processing easier i.e., products purification (Chen, Y. et al., 2003). For example, *Monascus* has been successfully cultured submerged condition for pigment production and versatile substrates like breads, rice and other amylaceous (starch, dextrins, glucose, maltose and fructose) materials for high productivity of red pigments which occurs due to glucose and maltose utilization.

It is reported that carotenoid from *Aspergillus sp.*, melanin type pigment from *Aspergillus niger*, water soluble red pigment from *Penicillium purpurogenum* are some investigated studies from fungi in submerged state fermentation (Malik, K. et al., 2012). Furthermore, many bacterial pigments too successfully produced using this technique (Nagpal, N. et al., 2011). Various commercial scale microorganism origin-based pigments such as prodigiosin, monascorubramie, astaxanthin, canthaxanthin, beta-carotene etc. produced using this fermentation method. Under submerged conditions, factors like nitrogen, pH temperature majorly affect the pigment production (Eyal and Spencer, 1992, Storebakken, T. et al., 1987).

Solid- state fermentation (SSF)

Solid state fermentation is defined as fermentation process performed on a non-soluble substance that acts both as physical support and source of nutrients in absence of free-flowing liquid (Chen, Y. et al.,2003). Solid state fermentation generally uses agro-industrial waste substrates, like bran, bagasse, and paper pulp etc. The main reason for utilization of agro-industrial waste as a substrate because it provides nutrient-rich solid waste materials which is quickly recycled In this fermentation technique, the substrates breakdown is very slow and continuous, so the same substrate can be utilized for long period of fermentation which help in controlled leakage of nutrients also. The main advantages of solid-state fermentation are that it is suitable for both fungi and microorganism require less moisture content for their growth, hence less water activity is suitable for this kind of fermentation techniques (Couto and Sanromán, 2006). Many pigments were reported using this technique and among all *Monascus* pigments lead with many reports. For example, red rice, a traditional Asian food, has for thousands of years been prepared by fermentation of steamed rice by *Monascus purpureus* variety in Solid state fermentaion. The fungus produces six different polyketide pigments colored from bright yellow to deep red, which have found applications both as food additives and pharmaceuticals (Malik, K. et al., 2012). For significant amounts of pigments by molds, substrates like rice, cassava or corn, wheat, oat and barley were used as substrates (Couto and Sanromán, 2006). *Aspergillus* sp. has been grown successfully on starch medium using different carbon sources wherein dextrin and maltose produced the highest color intensity. Some 38 significant studies include carotenoid by *Penicillium sp.*, riboflavin, a yellow pigment from *Eremothecium ashbyii* and *Ashbya gossypi*, beta-carotene by *Dunaliella* and *D. bardawil*, astaxanthin by *Haematococcus pluvialis*, phycocianin by *Spirulina sp.* etc (Rodríguez-Sáiz, M et al.,2010).

Low-cost substrates

Most of the time agro-industrial waste as by-products or extra, they cause environmental hazardous problem that's why now-a-days these cheaper rich carbohydrates agro-industrial waste is utilized as a substrate for various microbial fermentation. They decrease initial biological oxygen demand during microbial fermentation during the production of different biochemical compounds like colorant and pigments which are utilized in companies like pharmaceuticals, chemical and food (Tinoi, J. et al., 2006). These cheaper agro-industrial waste and their residues have potential being utilized by diverse group of microorganisms and produces beta carotene from citrus peel waste utilization, carotenoids production by whey ultrafiltrate, riboflavin from concentrated rectified grape must and astaxanthin from grape juice were reported (Surwase, S.N. et al., 2013 and Frengova, G. et al., 1994). High capital cost is involved for the treatment of these agro-industrial waste. The conversion of these waste in value added product material would help in economic gain and reduce environmental pollutant (Surwase, S.N. et al.,2013). During the selection agro-industrial waste, two factor play a key role for example carbon and nitrogen percentage because they are utilized by microorganisms and pigment production gets affected. Pigments produced on several agro-industrial wastes were listed down as shown in Table 2

Table 2: Different microorganisms and various inexpensive substrates used for pigments production

Substrate	Microorganism	Pigment type	References
Whey	*R. glutinis*	β -carotene	Surwase, S.N. et al.,2013
Potato medium	*R. mucilaginosa*	β -carotene	Frengova, G. et al., 1994
Crude glycerol	*R. glutinis*	carotenoids	Kim, S. et al., 2007
Chicken feather	*R. glutinis*	carotenoids	Martin, A.M. et al., 1993
Fermented radish brine	*R. glutinis*	β -carotene	Kim, S. et al., 2007
Mung bean waste flour and sweet potato extract	*R. glutinis*	carotenoids	Chen, Y. et al., 2003

Substrate	Microorganism	Pigment type	References
Mustard waste	*X. dendrorhous*	Astaxanthin	Tinoi, J. et al., 2006
Plant extracts	*X. dendrorhous*	Astaxanthin	Meyer and Preez, 1994
Coconut milk	*X. dendrorhous*	Astaxanthin	Marova, I. et al., 2012
Enzymatic hydrolysates of prehydrolysed wood	*X. dendrorhous*	Astaxanthin	Saenge, C et al., 2011
Sugarcane waste	*Streptomyces sp.*	melanin	Chen, Y. et al., 2003

Benefits and Applications of Microbial Pigments

Pigments produced by microorganisms are of traditional use in oriental countries and have been a subject of intense research in the present decades because of its potential for applications. Use of microorganisms and biotechnology would offer solutions to the problems of various industries especially food colorant industry. Fermentative production of colorants has a number of benefits that include: cheaper production, readily available raw materials, high yields and no seasonal variations (Couto and Sanromán, 2006). In contrast to higher plants, single cell algae and fungi are more appropriate for biotechnological production because they can be grown using existing culture techniques. Many fungi produce pigments which have application in both textile and food industrics (Tinoi, J. et al., 2006) Fungal pigments are routinely utilized as colorants for both foodstuffs and materials. Some of the more importantly utilized fungal pigments come from the water-soluble orange/red pigments produced by *Monascus sp.,* frequently used in rice wines in eastern countries (Surwase, S.N. et al., 2013). Pigments produced from *Monascus purpureus* Piedallu are used in wool dying, and an anthraquinone pigment obtained from *Penicillium oxalicum* (Frengova, G. et al.,1994). Mycelial extracts of some promising mushrooms are *Chroogomplus vinicolor* (which gives red tints), *Bankera violascens* (which gives greens) and *Collybia iocephala* (which gives blues), they have a remarkable potential for dyeing wool and silk fabrics. Carotenoids such as β-carotene and lycopene have been known to be produced by fungal cell factories. For example, β-carotene by *Blakeslea trispora, Mucor circinelloides*, and *Phycomyces blakesleeanus* and lycopene from *Fusarium sporotrichioides*, *Blakeslea trispora* are already in use as for food colorants (Martin, A.M. et al., 1994). Vitamin, riboflavin (vitamin B2) is a yellow food colorant that is fermentatively produced by the fungi *Eremothecium ashbyii* and *Ashbya gossypi* and has huge commercial potential (Kim, S. et al., 2007).

Microalgae such as *Haematococcus* (astaxanthin), *Spirulina* (phycocyanin), and *Dunaliella* (ß-carotene and other carotenoids) are already explored as sources of natural food colorants with some limitations of low productivity and contamination in the open culture system, where they are grown. Newly, a blue-green polyphenolic antioxidant pigment termed marennine from the Diatom *Haslea ostrearia* has been reported. This may aid as a promising food coloring additive (Saenge, C et al., 2011 and Marova, I. et al., 2012).

References

Aghaie-Khouzani, M. et al. Decolorization of some synthetic dyes using optimized culture broth of laccase producing ascomycete *Paraconiothyrium variabile*. Biochemical Engineering Journal, vol. 60, pp. 9-15, 2012.

Ahmad, W. A. et al. Application of Bacterial Pigments as Colorant: Springer, 2012.

Asker, D., and Ohta, Y. Production of canthaxanthin by extremely halophilic bacteria. Journal of bioscience and bioengineering, vol. 88, pp. 617-621, 1999.

Babitha, S. (2009). Microbial pigments. Biotechnology for agro-industrial residues utilisation, ed: Springer, pp. 147-162, 2009.

Babitha, S. et al., Jackfruit Seed--A novel substrate for the production of *Monascus* pigments through solid-state fermentation. Food Technology & Biotechnology, vol. 44, 2006.

Bechtold, T and Mussak, R. Handbook of natural colorants: John Wiley & Sons, 2009.

Boonyapranai, K. et al., Optimization of submerged culture for the production of naphthoquinones pigment by *Fusarium verticillioides*. Chiang Mai Journal of Science, vol. 35, pp. 457-466, 2008.

Buzzini, P and Martini, A. Production of carotenoids by strains of *Rhodotorula glutinis* cultured in raw materials of agro-industrial origin. Bioresource Technology, vol. 71, pp. 41-44, 2000.

Chen, Y. et al., Screening and characterization of astaxanthin-hyperproducing mutants of *Haematococcus pluvialis*. Biotechnology letters, vol. 25, pp. 527-529, 2003.

Couto, S.R. and Sanromán, M.A. Application of solid-state fermentation to food industry—a review. Journal of Food Engineering, vol. 76, pp. 291-302, 2006.

Cristea, D and Vilarem, G. Improving light fastness of natural dyes on cotton yarn, Dyes and Pigments, vol. 70, pp. 238-245, 2006.

Delgado-Vargas, F., Jiménez-Aparicio, A.R., Paredes-Lopez, O. Natural pigments: carotenoids, anthocyanins, and betalains-characteristics, biosynthesis, processing, and stability," Critical reviews in food science and nutrition, vol. 40, pp. 173-289, 2000.

Dufossé, L. et al., Microorganisms and microalgae as sources of pigments for food use: a scientific oddity or an industrial reality? Trends in Food Science & Technology, vol. 16, pp. 389-406, 2005.

Dufossé, L. Microbial Production of Food Grade Pigments. Food Technology & Biotechnology, vol. 44, 2006.

Durán, N. et al., Ecological-friendly pigments from fungi. Critical reviews in food science and nutrition, vol. 42, pp. 53-66, 2002.

Eyal, J. and Spencer, M. G. *Morchella rotunda sp.* useful for producing natural blue pigment. ed: Google Patents, 1992.

F. Delgado-Vargas and O. Paredes-López, Natural colorants for food and nutraceutical uses: CRC Press, 2002.

Freedman, D.M. Reasonable Certainty of No Harm: Reviving the Safety Standard for Food Additives, Color Additives, and Animal Drugs. Ecology LQ, vol. 7, p. 245, 1978.

Frengova, G. et al., Formation of carotenoids by *Rhodotorula glutinis* in whey ultrafiltrate. Biotechnology and bioengineering, vol. 44, pp. 888-894, 1994.

Hobson D.K. and Wales, D. S. Green'dyes. Journal of the Society of Dyers and Colourists, vol. 114, pp. 42-44, 1998.

Johnson, E.A. and Lewis, M. J. Astaxanthin formation by the yeast *Phaffia rhodozyma*. Journal of General Microbiology, vol. 115, pp. 173-183, 1979.

Joshi, V. et al., "Microbial pigments," Indian Journal of Biotechnology, vol. 2, pp. 362-369, 2003.

Kim, S. et al., Increased carotenoid production in *Xanthophyllomyces dendrorhous* G276 using plant extracts. Journal of microbiology-Seoul-, vol. 45, p. 128, 2007.

L. Dufossé, Microbial Production of Food Grade Pigments. Food Technology & Biotechnology, vol. 44, 2006.

Malik, K. et al., Microbial pigments: A review. Int. J. Microbial. Resour. Technol, vol. 1, pp. 361-365, 2012.

Marova, I. et al., Use of several waste substrates for carotenoid-rich yeast biomass production. Journal of Environmental Management, vol. 95, pp. S338-S342, 2012.

Martin, A.M. et al., Growth parameters for the yeast *Rhodotorula rubra* grown in peat extracts. Journal of fermentation and bioengineering, vol. 76, pp. 321-325, 1993.

Mata-Gómez, L.C. et al., "Biotechnological production of carotenoids by yeasts: an overview," Microbial Cell Factories, vol. 13, p. 12, 2014.

Meyer P. and Preez, J.D. Astaxanthin production by a *Phaffia rhodozyma* mutant on grape juice. World journal of microbiology and biotechnology, vol. 10, pp. 178- 183, 1994.

Mortensen, A. Carotenoids and other pigments as natural colorants. Pure and Applied Chemistry, vol. 78, pp. 1477-1491, 2006.

Mussatto, S.I. et al. "Use of Agro-industrial wastes in solid-state fermentation processes," Industrial waste. Croatia: In Tech, pp. 121-40, 2012.

Myers, R.H. and Anderson-Cook, C.M. Response surface methodology: process and product optimization using designed experiments vol. 705: John Wiley & Sons, 2009.

Nagpal, N. et al., Microbial pigments with health benefits-A mini review. Trends in Biosciences, vol. 4, pp. 157-160, 2011.

Nelis, H. J. and Leenheer, A. D. Microbial sources of carotenoid pigments used in foods and feeds. Journal of Applied Microbiology, vol. 70, pp. 181-191, 1991.

Pantanella, F. et al., Violacein and biofilm production in *Janthinobacterium lividum*. Journal of Applied Microbiology, vol. 102, pp. 992-999, 2007.

Rodríguez-Sáiz, M et al., *Xanthophyllomyces dendrorhous* for the industrial production of astaxanthin, Applied Microbiology and Biotechnology, vol. 88, pp. 645-658, 2010.

Saenge, C et al. Potential use of oleaginous red yeast *Rhodotorula glutinis* for the bioconversion of crude glycerol from biodiesel plant to lipids and carotenoids. Process Biochemistry, vol. 46, pp. 210-218, 2011.

Sanjay, K. et al., Optimization of carotenoid production by *Aspergillus carbonarius* in submerged fermentation using a response surface methodology. International Journal of Food Engineering, vol. 3, 2007.

Sheehan, H. et al., Color and scent: how single genes influence pollinator attraction. Cold Spring Harbor symposia on quantitative biology, pp. 117-133, 2012.

Shirata, A. et al. Isolation of bacteria producing bluish-purple pigment and use for dyeing. Japan Agricultural Research Quarterly, vol. 34, pp. 131-140, 2000.

Sikorski, Z. E. Chemical and functional properties of food components: CRC Press, 2006.

Sikorski, Z.E. Chemical and functional properties of food components: CRC Press, 2006.

Socaciu, C. Food colorants: chemical and functional properties: CRC Press, 2007.

Soliev, A.B. et al., Bioactive pigments from marine bacteria: applications and physiological roles. Evidence-Based Complementary and Alternative Medicine, vol. 2011..

Stahmann, K.P. et al. Three biotechnical processes using Ashbya gossypii, Candida famata, or *Bacillus subtilis* compete with chemical riboflavin production. Applied Microbiology and Biotechnology, vol. 53, pp. 509-516, 2000.

Storebakken, T. et al., Carotenoids in diets for salmonids: IV. Pigmentation of Atlantic salmon with astaxanthin, astaxanthin dipalmitate and canthaxanthin. Aquaculture, vol. 65, pp. 279-292, 1987.

Surwase, S.N. et al. Optimization of melanin production by *Brevundimonas sp.* SGJ using response surface methodology, 3 Biotech, pp. 1-8, 2012.

Surwase, S.N. et al., Optimization of melanin production by *Brevundimonas sp.* SGJ using response surface methodology. 3 Biotech, vol. 3, pp. 187-194, 2013.

Tinoi, J. et al. Utilization of mustard waste isolates for improved production of astaxanthin by *Xanthophyllomyces dendrorhous*. Journal of Industrial Microbiology and Biotechnology, vol. 33, pp. 309-314, 2006.

Tinoi, J. et al., Simplex optimization of carotenoid production by *Rhodotorula glutinis* using hydrolyzed mung bean waste flour as substrate. Process Biochemistry, vol. 40, pp. 2551-2557, 2005.

Vasanthabharathi, V. et al., "Melanin production from marine *Streptomyces*," African Journal of Biotechnology, vol. 10, pp. 11224-11234, 2013.

Venil, C and Lakshmanaperumalsamy, P. An insightful overview on microbial pigment, prodigiosin. Electronic Journal of Biology, vol. 5, pp. 49-61, 2009.

Venil, C.K. et al., Bacterial pigments and their applications. Process Biochemistry, vol. 48, pp. 1065-1079, 2013.

5

Biocatalysts and Their Role in the Pharmaceutical Industry

Rafiq A. Rather[1], Sonia Thapa[2], Madhulika Bhagat[1]

[1]*School of Biotechnology, University of Jammu, Jammu and Kashmir*
[2]*Cancer Pharmacology Division, CSIR-Indian Institute of Integrative Medicine, Canal Road, Jammu, Jammu and Kashmir*

Abstract

In biocatalysis, isolated enzymes, enzyme mixtures, immobilized enzymes or microbes (enzymes that are still present within the living cells) are used to perform chemical transformation of organic compounds into novel compounds. Both in academia and across chemical and pharmaceutical industry, biocatalysts are used as isolated preparations or in whole cell format or as recombinant proteins produced in alternate host cells. Biocatalysts catalyze novel small molecule transformations that are usually difficult or impossible using traditional synthetic organic chemistry. Biocatalysts exhibit high chemoselectivity, regioselectivity, diastereoselectivity and enantioselectivity towards their substrates. As a result, the use of biocatalysts during chemoenzymatic synthesis minimizes the issues of unwanted side-reactions such as decomposition, isomerization, racemization and rearrangement. The process of biocatalysis has become an important tool in the synthesis of active pharmaceutical ingredients (APIs) and enantiopure compounds. Biocatalysis offers a more sustainable, efficient, and less polluting methods for the production of APIs and advanced pharmaceutical intermediates. Biocatalysts have received tremendous attention in pharmaceutical industry because they typically catalyze reactions under mild conditions, so they do not consume much energy or emit much greenhouse gas. Biocatalysts are currently employed for the development of products in different fields, such as pharmaceuticals or intermediates of their production (e.g., antibiotics, statins, and enantiopure building blocks), fine chemicals (e.g., amino acids and vitamins), and food ingredients (e.g., sweeteners, lipids, and nutraceuticals). However, it would be helpful for the pharmaceutical industry to take advantage of developments in biochemistry, molecular cloning, random and site-directed mutagenesis, advanced tools for enzyme discovery, enzyme-directed evolution, high-throughput laboratory evolution techniques for biocatalyst optimization, immobilization, and fermentation technology to gain access to a new class of biocatalysts (enzymes), cells and other microbes.

Introduction

Biocatalysts, that exist in the form of pure enzymes, enzyme mixtures and whole cells, are of key importance in pharmaceutical industry (Truppo 2017). Biocatalysts are increasingly been used in research and for manufacturing new products in pharmaceutical industry (Tao et al. 2009). The use of natural substances, like enzymes or whole cells, to catalyze chemical reactions is called as biocatalysis (Adams et al. 2019). The oldest example of biocatalysis known to mankind is the production of ethanol from molasses (black treacle), a byproduct of the crystallization of cane sugar from concentrated cane juice (Ahluwalia and Kidwai 2004). An enzyme called invertase catalyzes the conversion of sucrose into glucose and fructose, and finally zymase converts glucose and fructose into ethanol (VIETHT). Similarly, nitrile hydratases have been used to synthesize acrylamide on an industrial scale, and enzymes (such as subtilisin) have been added to detergents for over 30 years (Bott and Betzel 2013; Nagasawa and Yamada 1989; Von der Osten et al. 1993; Yamada et al. 1996). With a few exceptions, biocatalysts in the early 2000s were primarily used for the synthesis or resolution of optically active intermediates. Combined (Chemo)–enzymatic cascades have enabled multiple total synthesis activities over the last 2 decades, and the use of enzymes is becoming routine in pharmaceutical industry (García-Junceda et al. 2015). In the past, chemical synthesis of pharmaceutical intermediates and active pharmaceutical ingredients (APIs) were limited to a subset of enzymes, such as ketoreductases (KREDs) and lipases (Bezborodov et al. 2016). Nowadays, every year many novel enzyme catalyzed reactions are published in high-impact journals. However, many such published articles lack vital information required to understand and reproduce the results. Standards for Reporting Enzymology Data (STRENDA) initiative was recently started to provide guidelines and explanations on topics of crucial relevance for biocatalysis (Gardossi et al. 2010). STRENDA initiative typically provides guidelines on issues such as the selectivity, specificity, productivity and stability of biocatalysts, as well as on methodological problems related to reactions in multiphase systems (Gardossi et al. 2010; Swainston et al. 2018). Machine learning driven enzyme engineering has played a vital role in the development of efficient biocatalysts for biotechnology, biomedicine, and pharmaceuticals (Mazurenko et al. 2019). However, an increase in the speed of enzyme engineering is required to take advantage of the ever-increasing opportunities for industrial biocatalysis (Truppo 2017). The most important biocatalysts used in industry include microbes, lipases, amaylases, cellulases, nitrilases and proteases

(McAuliffe et al. 2007). Over the last two decades, biocatalysis-based technologies have become a promising area of research and has made the large scale preparation of "active pharmaceutical ingredients" (APIs) economical and environment-friendly (Busacca et al. 2011). By the definition of WHO, "active pharmaceutical ingredients" refers to any substance or combination of substances used in a finished pharmaceutical product (FPP), intended to furnish pharmacological activity or to otherwise have direct effect in the diagnosis, cure, mitigation, treatment or prevention of disease, or to have direct effect in restoring, correcting or modifying physiological functions in human beings. APIs are traditionally produced by pharmaceutical companies or multinational corporations. The use of biocatalysis is fundamental to green chemistry and reduces the overall carbon footprint of chemical manufacturing (Busacca et al. 2011; Hollmann and Fernandez-Lafuente 2021). Biocatalysis has garnered attention for its role in addressing global challenges such as climate change, substitution of fossil fuels, feeding a growing population (Mohan et al. 2016; Philp and Science 2015). Biocatalyst-mediated bioprocessing of food, feed, and nonfood industries improves resource efficiency and paves way to get more products out of the raw biomaterials (Chourasia et al. 2020; Fernandes 2010). A number of biocatalysts including proteases, lipases, carbohydrases, tannases, and phytases are used in the synthesis of bioactive compounds (Chourasia et al. 2020). In the present bioeconomy era, enzymes and biocatalysts are playing fundamental role in sustainable production of bio-based products (Mohan et al. 2016; Pellis et al. 2018; Yuryev and Liese 2010). With the help of biocatalysts, industrial processes have become more sustainable because biocatalytic processes are milder, use less energy, producing less waste water compared to non-enzymatic chemical processes (Patel 2000). Thus, the introduction of biocatalysis presents novel opportunities to develop 'green' synthetic methods to pharmaceuticals and other active pharmaceutical ingredients, because enzymes are typically functional in aqueous media and act under mild conditions of temperature, pH and ionic strength. Although, biocatalysts enhance the rate of a reaction, they do not affect its thermodynamics properties. It is now clear that biocatalysis is plays a significant role in the synthesis of chemicals, active pharmaceutical ingredients, pharmaceutical intermediates where high degree of chemoselectivity, regioselectivity and enantioselectivity is needed (Patel 2000). The chemical selectivity of biocatalysts is often divided into four discrete categories- substrate selectivity, stereoselectivity, regioselectivity and functional group selectivity (Rozzell and Chemistry 1999). Biocatalysts have substituted organocatalysis and metal catalysis in a variety of industries such as in processing of starch, production of biofuels, production of detergents and soaps, production of foods and beverages, and production of active pharmaceutical ingredients (Bisogno et al. 2017). Since biocatalysis involves the use of enzymes and whole cells (such as plants, fungi and bacteria), it has historically been classified into homogeneous catalysis and heterogeneous catalysis (Cornils et al. 2020; Yuryev and Liese 2010). Like chemocatalysis, biocatalysis can be homogeneous or heterogeneous. The modification of enzymes with certain polymers is emerging as a promising approach to finely tune the molecular interactions of biocatalysts in diverse solvent environments. In anhydrous ionic liquids, for instance, modification of lipase with poly (4-acryloylmorpholine) enhances solubility and transesterification activity (Chado et al. 2018). In industrial set up, biocatalysts are increasingly been used to achieve eco-friendliness, cost-effectiveness and long-term sustainability of advanced pharmaceutical ingredients (APIs). However, finding a suitable biocatalyst for any particular process fulfilling these criteria is still a challenging task and has to be taken care of for upscaling of the process.

Classification of biocatalysts

Biocatalysts are typically derived from plants, animals, fungi and microbes. However, microbes, in particular, are important sources of biocatalysts (Cano-Flores et al. 2020; Goswami and Stewart 2016). The advantages and disadvantages of different biocatalysts can vary depending on the source and biochemical composition of the biocatalyst. There are six major classes of biocatalysts based on the type of biochemical reaction they catalyze-oxidoreductases, transferases, hydrolases, lyases, isomerases, and ligases (Table I) (VIETHT). The biochemical processes where natural or modified enzymes are used to perform organic synthetic reactions are termed chemoenzymatic syntheses; while the reactions catalyzed by the enzyme are classified as chemoenzymatic reactions. Biocatalysts include pure enzymes, enzyme mixtures or micro-organisms that accelerate the rate of chemical reactions. Biocatalysts are believed to evolve through successive cycles of mutation, selection and amplification (Jäckel and Hilvert 2010). In laboratory settings, directed evolution (a cyclic process in which gene diversification alternates with screening for or selection of functional gene variants) is often used to improve and modify biocatalysts (Arnold and Volkov 1999). Compared to conventional chemical catalysts, biocatalysts offer a number of advantages. Biocatalysts, for example, are steroselective, enantioselective, regioselective and chemoselective (Mu et al. 2020). These properties make biocatalysts useful for synthesizing enantiopure compounds and for preparing intermediates and active pharmaceutical ingredients (APIs) for the pharmaceutical industry (Mu et al. 2020). The purpose of a pharmaceutical industry is the discovery, development, production, and marketing of new drugs or existing drugs designed to be administered (or self-administered) to patients with the aim of curing them, preventing them from contracting disease, or alleviating their symptoms (Taylor 2015). To achieve sustainability in pharmaceutical industry, it is imperative that we continue to explore and utilize enzymes and microorganisms for providing goods and services characterized by selectivity, specificity, productivity, and environmental friendliness to mankind (Wiltschi et al. 2020). However, Food and Drug Administration (FDA) approval is obligatory for new pharmaceutical products in the United States as they must be both safe and effective (Patel 2008). In the early 1990s, a technique called "*in vitro selection*" was developed which allowed scientists to isolate a great number of DNAzymes, which are single-stranded DNA molecules with catalytic activity (Huang and Liu 2020). Nevertheless, multiple 'RNA-cleaving DNAzymes'-based functional probes have been developed that are used in diagnostics, intracellular imaging, and *in vivo* therapeutics (Morrison et al. 2018). However, DNAzymes have a narrower range of catalytic functionality than protein enzymes, which limits their applicability in pharmaceutical industry (Morrison et al. 2018). In recent years, the use of biocatalysts to synthesize active pharmaceutical ingredients (APIs) has grown rapidly (Patel 2008). The above-mentioned advantages have contributed to the industrial adoption of biocatalysts for the synthesis of commercially interesting compounds, such as pharmaceuticals and fine chemicals, which are needed to satisfy economic requirements while respecting green chemistry requirements (Wu et al. 2021). Moreover, advances in biotechnology and immobilization techniques, as well as recent advances in protein engineering have improved the efficiency of biocatalysis in synthesis by making enzyme production cheaper and providing access to a large library of versatile biocatalysts.

Table 1: Classification of enzymes and the nature of reactions catalyzed

Class of enzymes	Nature of reaction catalyzed	Subclasses
Hydrolases	Hydrolases are a class of enzymes that break chemical bonds with water, resulting in the cleavage of larger molecules into smaller ones	amidases, esterases, glycosidases, lipases, nucleosidases, phosphatases, proteases, peptidases
Oxidoreductases	Oxidoreductases are enzymes that catalyze oxidation/reduction (redox) reactions	dehydrogenases, oxidases, oxygenases, peroxidases, reductases
Transferases	Transferases act by catalyzing the transfer a chemical group from one substance to another	aminotransferases, phosphotransferases, glycosyltransferases,
Lyases	Lyases catalyze the elimination reactons	Aldolases, decarboxylases, dehydratases, pectinases
Isomerases	Isomerases catalyze the isomerisation reactions	Epimerases, racemases

Class of enzymes	Nature of reaction catalyzed	Subclasses
Ligases	Ligases catalyze the synthesis of a covalent bond by joining two molecules together with the help of energy released by ATP hydrolysis	C-C ligases, C-N ligases, C-O ligases, C-S ligases

Whole-cell Biocatalysts

Microbes have evolved in response to a wide variety of substrates and environments, resulting in the production of numerous metabolites and biocatalysts (enzymes) (de Carvalho 2017). Biocatalysts are increasingly being used for the synthesis of some valuable added fine chemicals, bulk chemicals and active pharmaceutical ingredients (APIs), both in industrial and laboratory settings (Lin and Tao 2017). Biocatalysts include all types of enzymes, including natural enzymes, wild-type enzymes, mutant enzymes, recombinant enzymes, modified enzymes, engineered enzymes, groups of enzymes, and cell extracts and whole cells. Whole cell biocatalysts have several advantages which include high catalytic efficiency and high selectivity. The high selectivity may be chiral, positional, and functional group-specific (Lin and Tao 2017). The traditional methods for preparing biocatalysts are time-consuming and expose the catalyst to potentially harmful conditions (Abdelraheem et al. 2019). However, high-throughput experiments in the recent times have become increasingly important in identifying and isolating biocatalysts. Although humans have been using the biocatalysts for centuries, it is only during the recent times that we have been able to control their production, to develop new biocatalysts with high thermodynamic stability and productivity and to improve their yields under challenging reaction conditions. There are also certain challenges related to the implementation of biocatalysis in pharmaceutical manufacturing (Rosenthal et al. 2018). For instance, a newly identified biocatalyst does not have all the properties required for synthesis of active pharmaceutical ingredient (APIs), so further protein engineering is almost always required. Besides, there is a high attrition rate in pharmaceutical development, which means that the return on investment when developing a biocatalytic route is uncertain. Therefore, resource allocation must be phased as the probability of success increases as clinical milestones are reached (Devine et al. 2018; Rosenthal et al. 2018). Although many commercially available biocatalysts are available, they catalyze limited reaction classes given the diversity of substrates and reactions of interest to pharmaceutical industry (Adams et al. 2019). However, DNA sequencing technologies and metagenomic methods have increased the sequence space known, and cheaper DNA synthesis technologies are enabling the development of diverse set of new enzymes for use in pharmaceutical industry (Schmeisser et al. 2007). However, putting all of this new diversity in the hands of a chemist within a timeframe that allows API development is still a challenge. In addition, there are commercial suppliers for some of these enzymes, such as lipases, but these supplies are not sufficient to meet manufacturing needs. Biocatalysis have replaced chemical catalysis in many fields and the production of pharmaceutical intermediates and APIs by biocatalysts is quite common these days (Table II). Halohydrin dehalogenase (HHDH) from *Agrobacterium radiobacter* AD1 is used for the industrial manufacturing of ethyl (R)-4-cyano-3-hydroxybutanoate (HN), a key chiral synthon of a cholesterol-lowering drug of atorvastatin (Luo et al. 2017). However, it is difficult to discover and characterize new and better-performing biocatalysts since majority (~99%) of the microbes present in many environments cannot cultured/grown in the laboratory settings. Extremophilic marine microorganisms found in hot springs, seafloors, hydrothermal vents, and seafloors are important sources of extremozymes, such as cellulases, amylases, proteases and dehydrogenases (Van Den Burg 2003). Extremozymes are potential source for industrial applications (Dumorné et al. 2017). *Alcanivorax dieselolei* B-5(T), for example, is an important source of leucine dehydrogenase, an enzyme that produces L-tert-leucine from trimethylpyruvic acid. L-tert-leucine is an important component of many chiral drugs.

Table 2: Production of pharmaceuticals by whole-cell biocatalysts

Compound	Microorganism	Reference
Ethyl (R)-4-cyano-3-hydroxybutanoate (HN)	*Agrobacterium radiobacter* AD1	(Luo et al. 2017)
Acrylamide	*Rhodococcu rhodochrous J; Rhodococcus rhodochrous PA-34*	(Lan et al. 2017; Raj et al. 2008)
Nicotinamide	*Rhodococcu rhodochrous J1*	(Lan et al. 2017)
(S)-ibuprofen	*Acinetobacter sp. AK226*	(Asano and Kaul 2012)
L-Carnitine	*Escherichia coli*; *Proteus* sp.; *Agrobacterium*, *Rhizobium* HK1349	(Bernal et al. 2007; Stephanopoulos 1996)
Violacein	*C. violaceum; Duganella sp.; Pseudoalteromonas sp.; Iodobacter sp.; Massilia sp.*	(Park et al. 2021)
Vitamin B12	*P. denitrificans; Propionibacterium shermanii*	(Survase et al. 2006)
L-lysine	*Corynebacterium glutamicum*	(Gopinath et al. 2012; Wang et al. 2019a)
Glutarate	*Escherichia coli* strains BL21-22AB and BL21-YDT	(Wang et al. 2019a)

Compound	Microorganism	Reference
L-glutamate	*Corynebacterium glutamicum*	(Gopinath et al. 2012)
Ethyl secodione	*Ralstonia* sp.	(Chen et al. 2019)
(R)-10-hydroxy stearic acid	*Elizabethkingia meningoseptica; Rhodococcus erythropolis*	(Engleder et al. 2017; Lorenzen et al. 2018)
Dodecane diacid	*Yarrowia lipolytica* strains; recomb. *Candida tropicalis*	(Mokgoro 2003; Wu et al. 2021)
(S)-duloxetine	*Rhodosporidium toruloides*	(Chen et al. 2016; Shahzad et al. 2019)
(R)-mandelic acid	*Pseudomonas putida* MTCC 5110	(Asano and Kaul 2012)

Mature Reaction Classes and Biocatalysts

A variety of reaction classes have been identified where biocatalysts may play a role in pharmaceutical industry, but remain underdeveloped for a variety of reasons (Schoemaker et al. 2003). Like all catalysts, biocatalysts enhance the rate of a reaction by lowering the activation energy needed for the reaction to occur. There are four major types of reactions carried out by biocatalysts that are important in health-care industry.

Amide Synthesis

Amide bonds are the most common structural motifs in active pharmaceutical ingredients (APIs) with unparalleled utility in drug discovery and drug design (Goswami and Van Lanen 2015). The formation of these chemical bonds has been the subject of considerable research. In a typical process, the ester is first hydrolyzed to give a carboxylic acid. The carboxylic acid is then activated by changing its structure to an acyl chloride, an acyl anhydride, or another functional group so that it is more prone to nucleophilic attack by an amine (Table III; Figure 1). Biocatalysis is considered a viable method for the formation of amide bonds (Goswami and Van Lanen 2015). An important example of biocatalyst that causes aminolysis of methyl esters into amides is *Pseudomonas stutzeri* lipase (Table III; Figure 2)(Van Pelt et al. 2011). It has been reported that this biocatalysts works better on bulky methyl esters and on secondary amines while another biocatalyst (Biocatalyst CAL-B) works better with methyl 2-phenylpropionic as the acyl donor (Adams et al. 2019). The use of CAL-B in the selective direct fluorination–amidase strategy led to the sustainable synthesis of enantiopure fluorolactam derivatives. Following pH adjustment, it was determined that dimethyl 2-[3-(chloroamino)propyl]-2-fluoromalonate was successfully converted (through the use of CAL-B) to its enantiopure lactam with a 47% isolated yield and 98% ee (S) (Table III; Figure 3) (Willis et al. 2016). SpL, a lipase found in Sphingomonas sp. HXN-

200, forms amide bonds using variety of methyl esters and carboxylic acids as substrates (Table III; Figure 4) (Hill and Sutherland 2016).

Transaminases

The synthesis of Sitagliptin, an anti-type 2 diabetic drug, is an excellent example of the use of engineered transaminases (TAs) on an industrial scale (Alcántara 2018). *Roseomonas deserti* (TARO) has been found to be the best source of transaminase for synthesis of this drug (Khobragade et al. 2021). Transaminases are regarded as the biocatalyst of choice for the synthesis of chiral primary amines (Gomm and O'Reilly 2018). ω-Transaminases (ω-TAMs) have been extensively used for the synthesis of chiral molecules, providing a possible "green alternative" to overcome the limitations of conventional chemical methods (Kelly et al. 2018). Likewise, the engineering of an amine transaminase has allowed the efficient synthesis of the chiral sacubitril, (2R,4S)-5-([1,1′-biphenyl]-4-yl)-4-amino-2-methylpentanoic acid, which is an important component of Entresto (Novick et al. 2021). These days many commercial panels of transaminases are available which are used to provide off the shelf solutions to challenging problems. A stereospecific solid-phase screening assay is often used to identify (R)- and (S)-selective ω-aminotransferases (Willies et al. 2016). However, transaminase mediated reactions are reversible in nature (Table III; Figure 5) which thermodynamically limits the use of TAs in the synthesis of APIs. The thermodynamic limitation of these reactions can be overcome by using *in-situ* product removal as a technique for driving the transaminase reaction in the direction of product formation.

Ketoreductases/Alcohol Dehydrogenases

Ketone reductases (KREDs, also known as alcohol dehydrogenases) are used as biocatalysts for the synthesis of pharmaceutical ingredients, and are often combined with hydrolases (Micklitsch et al. 2019). KREDs facilitate the reversible transfer of a hydride (H^-) from reduced nicotinamide adenine dinucleotide phosphate (NADPH) to aldehydes and ketones. These enzymes are highly stereospecific and act on a wide range of substrates (Table III; Figure 6). The use of KREDs is environment-friendly methodology and has proved very useful for the preparation of chiral alcohols compared to traditional chemical catalysis (Micklitsch et al. 2019). Chiral alcohols are fundamental structural and functional motifs in many pharmaceuticals and fine chemicals (Wang et al. 2019b). Chiral hydroxyl moieties are key intermediates of multiple APIs of drug candidates. Dulox alcohol (S)-5 is a key precursor for the antidepressant duloxetine. Likewise, ketoreductase have also been found to catalyze the stereoselective bioreduction of α-nitro ketones into their corresponding β-nitro alcohols (Wang et al. 2019b). (R)-tembamide, (S)-tembamide, (S)-toliprolol, (S)-moprolol and (S)-propanolol are key examples of bioactive molecules containing amino alcohol moiety (Wang et al. 2019b).

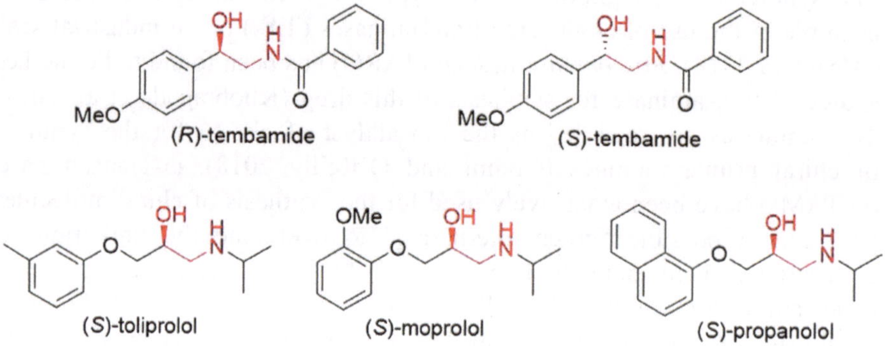

(*R*)-tembamide (*S*)-tembamide

(*S*)-toliprolol (*S*)-moprolol (*S*)-propanolol

Fig. 7: Bioactive molecules containing vicinal amino alcohol moieties (Wang et al.2019)

Nitrilases

Nitrilases are biocatalysts that hydrolyze nitriles to their corresponding carboxylic acids and ammonia (Shen et al. 2021). Due to their selectivity and high activity under mild reaction conditions, nitrilases are of great interest to the chemical and pharmaceutical industries (Gong et al. 2012a). Further, nitrilases exhibit regio-, diastereo-, chemo- and enantioselective properties. In some situations, nitrilase catalyzed nitrile re-racemisation (e.g., of hydroxynitriles) allows potentially complete (100%) enantioselective transformations. The success of nitrilase in manfacturing nicotinic acid, (R)-(-)-mandelic acid and acrylic acid is proof of the great potential of nitrilase in industry (Gong et al. 2012b; Nagasawa et al. 1990). In the pharmaceutical industry, nitrilases have been useful in the synthesis of pregabalin and in the synthesis of (R)-4-cyano-3-hydroxybutyrate, which is a potential intermediate for synthesis of Atorvastatin (Patel 2008). Nitrilases have been used to resolve rac-3-oxocyclohexane-1-carbonitrile, an important drug precursor (Table III; Figure 8) (Adams et al. 2019). For example, a recently identified nitrilase enzyme NIT-G6 enabled a 20 g scale reaction at 50 gL^{-1} substrate loading which provided (S)-3-oxocyclohexane-1-carbonitrile with an isolated yield of 28% and >99% ee (Adams et al. 2019). The product resulting from the enzymatic route enabled process chemists to develop new simpler strategies to prepare the API. The high enantioselectivity, and reaction robustness supports the industrial applicability of these enzymes.

Table 3: Major reaction classes and representatives reaction schemes

Mature reaction classes	Reaction scheme
Amide Synthesis	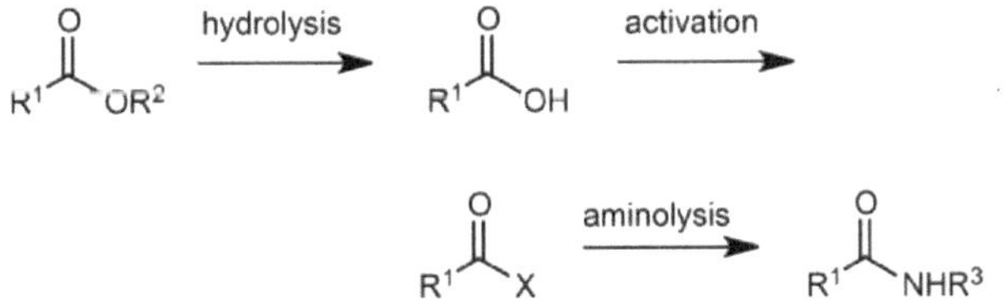 **Fig. 1:** Amide bond synthesis from esters
	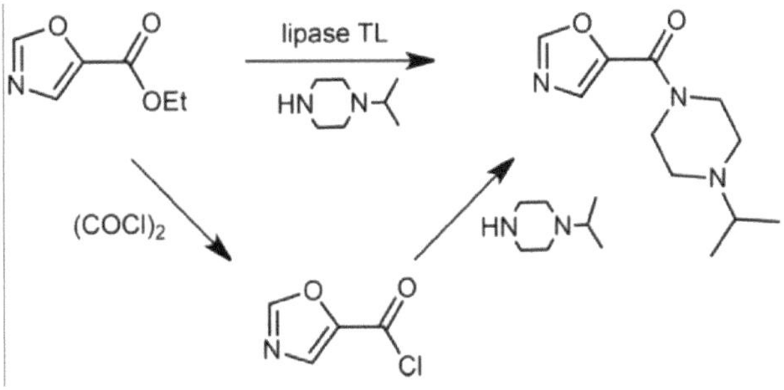 **Fig. 2:** Lipase TL-catalysed synthesis of **(4-isopropylpiperazin-1-yl)(oxazol-5-yl)methanone**
	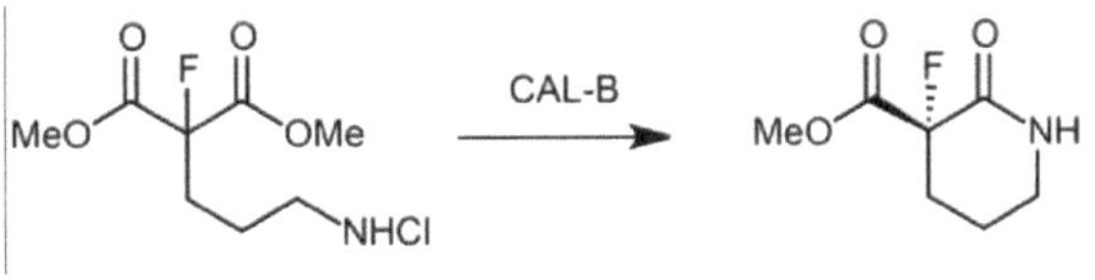 **Fig. 3**: CAL-B-catalysed synthesis of methyl (S)-3-fluoro-2-oxopiperidine-3-carboxylate
	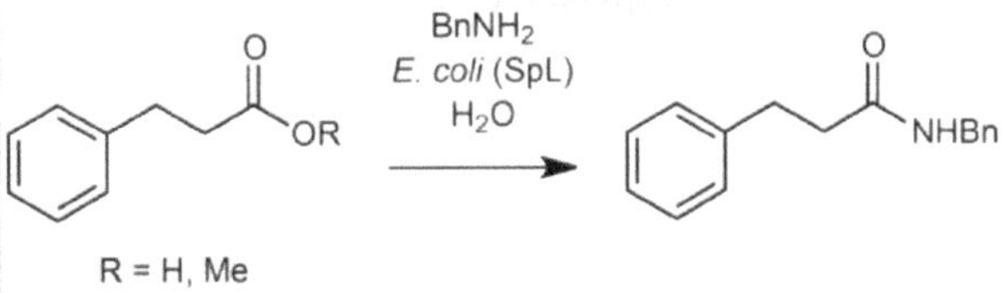 **Fig. 4**: SpL-catalysed amidation of carboxylic acid and methyl ester
Transamination and transaminases	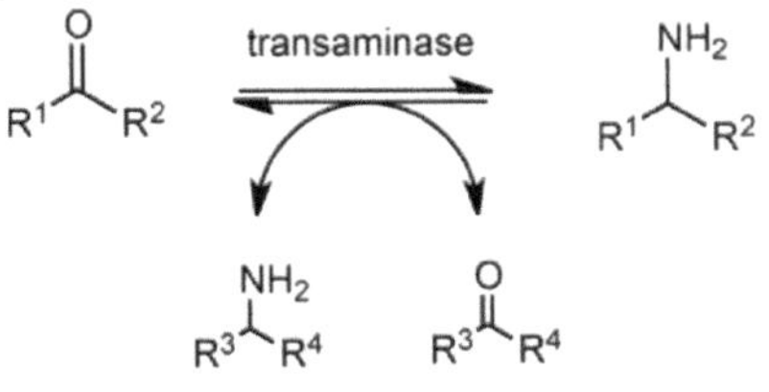 **Fig. 5**: A typical amino group transfer reaction catalyzed by transaminases

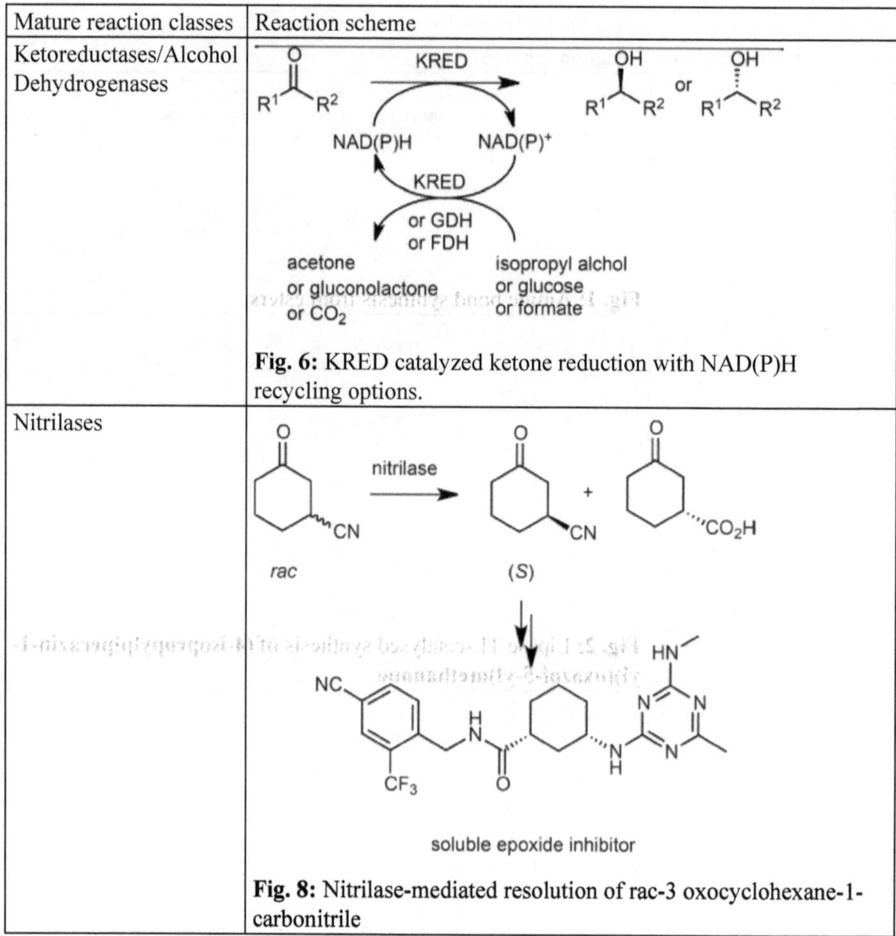

Mature reaction classes	Reaction scheme
Ketoreductases/Alcohol Dehydrogenases	**Fig. 6:** KRED catalyzed ketone reduction with NAD(P)H recycling options.
Nitrilases	**Fig. 8:** Nitrilase-mediated resolution of rac-3 oxocyclohexane-1-carbonitrile

Factors that Enable Biocatalysis

Several factors influence biocatalysis, such as reaction design, biocatalyst selection and optimization, and bioprocess development (Bell et al. 2021). For biocatalysis to be successful, however, the biocatalytic process must be intensified. This can be achieved through reactor miniaturization, continuous operation, *in situ* product removal, process analytics and cascade reactions involving fewer steps (Žnidaršič-Plazl and Chemistry 2021). Biocatalysis has been used for the synthesis of a wide range of chemicals and pharmaceuticals, including materials and polymers (e.g., acrylamide), fragrances and flavours (e.g., ambrox, fragrance intermediate; nootkatone), pharmaceuticals (e.g., islatravir, penicillin G), fine chemicals (e.g., vanillin), biopharmaceuticals/ new modalities (e.g., antibody–drug conjugates) and food (e.g., aspartame) (Bell et al. 2021). Throughout successful biocatalysis, these development stages are interdependent and must be integrated (Van Roon et al. 2007). A single or multistep biocatalytic process is typically designed from a target molecule, using expertise and precedent literature from organic synthesis and biocatalysis. Biocatalyst databases such as RetroBioCat are used to accelerate this stage and facilitate de novo biosynthetic design (Finnigan et al. 2020). RetroBioCat is a computer-aided synthesis planning tool for biocatalytic cascades and reactions (Finnigan et al. 2021). Similarly, YASARA is another tool that is used to acquire structural guidance regarding enzymes and their substrate interactions (Land and Humble 2018). As soon as a biocatalytic process has been designed, suitable enzymes are selected for each step and tested. In the next step, a fast optimization of catalytic activity can be achieved by using advanced tools in protein engineering, directed evolution strategies, and computational design (Barrozo et al. 2012; Hibbert and Dalby 2005). Subsequently, strict reaction conditions (such as pH, temperature, solvents etc) are set at reasonable timescales for successful bioprocess development. By enhancing the stability, activity, enzyme-substrate affinity, and thermo-stability of biocatalysts, a greater yield can be expected from biocatalysis. By optimizing the safe development of processes, we can ensure both the quality and productivity of the drugs produced.

Applications of Biocatalysts in Pharmaceutical Industry

Biocatalysis has been profitably used for the production of pharmaceutically active ingredients and therapeutically important drugs. The typical examples of biocatalytically produced drugs are given below:

Sitagliptin

Sitagliptin is a dipeptidyl-peptidase-4 (DPP4) inhibitor that was successfully produced via biocatalysis as an anti-type 2 diabetic drug (Ahren et al. 2007; Patel 2011). Genetically engineered R-stereospecific transaminases (R-ATA, ATA-117) from *Arthrobacter* species have been successfully used in the production of this drug in a highly stereospecific manner (Guan et al. 2015). The R- and S- stereospecific ATAs are also used in the formation of wide range of drugs such as niraparib and the production of an antagonist of orexin receptor (Ferrandi et al. 2018).

Boceprevir

This drug is produced via chiral amine synthesis approach and is used for the treatment of chronic hepatitis C infections. The production of this drug involves the asymmetric amine oxidation of bicyclic proline intermediate with the help of monoamine oxidase (MAO) from fungus *Aspergillus niger*

(Ghislieri et al. 2013; Li et al. 2012). Genetically engineered monoamine oxidases (MAOs) are also used in the formation of telaprevir (another hepatitis C drug) and various other synthetic drugs such as solifenacin, levocetirizine, (R)-harmicine along with few natural alkaloid products (confine, harmicine, elegance, and leptaflorine) (Ghislieri et al. 2018; Ghislieri et al. 2013).

Montelukast

Montelukast (also known as Singulair) is an anti-asthmatic drug that is produced by an engineered keto-reductase (KRED) from bacterium *Lactobacillus kefir* which exhibits significant enantioselectivity (99.9%) and stays stable in 70% organic solvent and temperature of 45°C (Choi et al. 2015; Liu et al. 2019). Several other drugs such as atorvastatin, crizotinib, duloxetine, and phenylephrine have also been developed by biocatalytic process using KRED from bacterium *Lactobacillus kefir* (Liu et al. 2019).

Atorvastatin

Atorvastatin (brand name lipitor) belongs to a family of drugs known as 'statins'. This drug functions by blocking the biosynthesis of low density lipoprotein (LDL)-cholesterol in the liver via inhibition of 3-hydroxy-3-methylglutaryl-coenzyme A (HMG-CoA) reductase. The synthesis of the different intermediates of this drug is carried out by a cascade of biocatalysts such as alcohol dehydrogenase, 2-deoxy-d-ribose 5-phosphate aldolase, nitrilase, lipase, glucose dehydrogenase (GDH), halohydrin dehalogenase (HHDH) and KRED. KRED is needed for the formation of hydroxy nitrile, an important intermediate in the formation of atorvastatin.

Pregabalin

Pregabalin is an alkylated analog of lipophilic γ-aminobutyric acid (GABA) that is used as an anticonvulsant drug to treat various neurological disorders, such as epilepsy, anxiety, nerve pain and fibromyalgia (Ciaramella 2019; Froestl 2011). This drug is synthesized by lipolase-mediated resolution of rac-2-carboxyethyl-3-cyano-5-methylhexanoic acid ethyl ester (CNDE) to (S)-2-carboxyethyl-3-cyano-5-methylhexanoic acid. Using heat, 2-carboxyethyl-3-cyano-5-methylhexanoic acid is decarboxylated resulting in the formation of (S)-3-cyano-5-methylhexanoic acid ethyl ester, which is an important chiral intermedaite of pregabalin. Lately, a novel esterase EstZF172 was discovered in *Pseudomonas sp.* that acts as a biocatalyst of choice for the efficient synthesis of a chiral intermediate of pregabalin (Xu et al. 2015). Similarly, a newly isolated *Morgarella morganii* ZJB-09203 has been use for biocatalytic synthesis of chiral intermediate of pregabalin with high substrate loading(Zheng et al. 2013). The latest data suggests that pregabalin interacts with α2δ1 and α2δ2 subunits of voltage-gated calcium channels in the CNS with high affinity (Li et al. 2011). This chemoenzymatic process makes the synthesis of this drug environmental friendly by reducing organic solvent usage.

7-aminocephalosporic acid (7-ACA)

Cephalosporins are a group of semi-synthetic β-lactam antibiotics that interfere with peptidoglycan synthesis in bacterial cell walls (Marshall and Blair 1999). Cephalosporins are synthesized from 7-aminocephalosporanic acid (7-ACA) which in turn is produced by the enzymatic deacylation of cephalosporin-C (CPC). In a two-step enzymatic conversion of CPC to 7-ACA, D-amino acid oxidase (DAAO) and 7-β-(4-carboxybutanamido)-cephalosporanic acid acylase (GLA) are both used (Conlon et al. 1995). Cephalosporin C acylase from microbes has also been used to convert CPC to 7-ACA in a single step(Hardianto et al. 2016). Likewise, the synthesis of semisynthetic penicillins with 6-aminopencillanic acid has also been reported using penicillin acylase(Duggleby et al. 1995). Immobilized penicillin acylase is a biocatalyst of choice for the kinetically controlled industrial production of semi-synthetic antibiotics (amoxicillin and ampicillin) in aqueous media (Illanes and Fajardo 2001). Amoxicillin and ampicillin are prepared by condensing 6-aminopenicillanic acid with the amide or ester of d-(−)-4-hydroxyphenylglycine and d-(−)-phenylglycine respectively (Basso and Serban 2020). In the same way, the cephalosporin antibiotics cefadroxil and cephalexin can be prepared from 7-aminodesacetoxycephalosporanic acid.

Conclusions

Biocatalysis has made an incredible journey so far and is a successful technique for the numerous biotransformation processes in the production of active pharmaceutical ingredients and drugs in pharmaceutical industries. Pharmaceutical industries and chemical industries have immensely benefited from the use of biocatalyst-based technologies. However, the successful development of biocatalysis that are economically feasible and sustainable depends on enzymes that can function under extreme reaction conditions regardless of their chemical composition. Several recent studies have indicated that most of the biocatalysts are being used in the synthesis of pharmaceuticals or drugs or intermediates to replace chemical catalysts, but there are concerns about their stability, selectivity, and specificity. There are a number of tools and techniques available for enhancing the stability, activity, enzyme-substrate affinity, and thermostability of biocatalysis with an expectation to increase the yield. For example, by incorporating artificial metabolic pathways, cell factory design, and nanotechnology approaches, suitable biocatalytic processes can be further improved to enhance the yield. Through optimized safe process design, the production of pharmaceutical intermediates, APIs and drugs will be of high quality and productivity. In summary, biocatalysis provides a safe, inexpensive, and sustainable way to produce high-value fine chemicals, APIs and other pharmaceutical drugs. However, it still presents a number of exciting challenges for potential future development and warrants extensive research.

Author Contributions

RAR conceived the idea and wrote the chapter with the assistance from ST and MB. All authors contributed to the chapter and approved the submitted version.

Acknowledgments

The corresponding author is grateful to SERB-National Postdoctoral Fellowship program and Science and Engineering Research Board, Department of Science and Technology, New Delhi, India (Grant Number: PDF/2016/002730) for financial support.

References

Abdelraheem EM, Busch H, Hanefeld U, Tonin FJRC, Engineering (2019) Biocatalysis explained: from pharmaceutical to bulk chemical production 4:1878-1894

Adams JP, Brown MJ, Diaz-Rodriguez A, Lloyd RC, Roiban GDJAS, Catalysis (2019) Biocatalysis: A pharma perspective 361:2421-2432

Ahluwalia V, Kidwai M (2004) Biocatalysts in Organic Synthesis. In: New Trends in Green Chemistry. Springer, pp 88-107

Ahren BJBP, Endocrinology RC, Metabolism (2007) DPP-4 inhibitors 21:517-533

Alcántara AJJMCDD (2018) Biotransformations in drug synthesis: a green and powerful tool for medicinal chemistry 1:1-7

Arnold FH, Volkov AAJCoicb (1999) Directed evolution of biocatalysts 3:54-59

Asano Y, Kaul P (2012) 7.7 Hydrolysis and Reverse Hydrolysis: Selective Nitrile Hydrolysis using Nitrilase and its Related Enzymes

Barrozo A, Borstnar R, Marloie G, Kamerlin SCLJIjoms (2012) Computational protein engineering: bridging the gap between rational design and laboratory evolution 13:12428-12460

Basso A, Serban S (2020) Overview of immobilized enzymes' applications in pharmaceutical, chemical, and food industry. In: Immobilization of Enzymes and Cells. Springer, pp 27-63

Bell EL et al. (2021) Biocatalysis 1:1-21

Bernal V, Sevilla Á, Cánovas M, Iborra JL (2007) Production of L-carnitine by secondary metabolism of bacteria Microbial Cell Factories 6:31 doi:10.1186/1475-2859-6-31

Bezborodov A, Zagustina NJAB, Microbiology (2016) Enzymatic biocatalysis in chemical synthesis of pharmaceuticals 52:237-249

Bisogno FR, López-Vidal MG, de Gonzalo GJAS, Catalysis (2017) Organocatalysis and Biocatalysis Hand in Hand: Combining Catalysts in One-Pot Procedures 359:2026-2049

Bott R, Betzel C (2013) Subtilisin enzymes: Practical protein engineering vol 379. Springer Science & Business Media,

Busacca CA, Fandrick DR, Song JJ, Senanayake CHJAS, Catalysis (2011) The growing impact of catalysis in the pharmaceutical industry 353:1825-1864

Cano-Flores A, Gómez J, Escalona-Torres IS, Velasco-Bejarano B (2020) Microorganisms as biocatalysts and enzyme sources. In: Microorganisms. IntechOpen,

Chado GR, Holland EN, Tice AK, Stoykovich MP, Kaar JLJB (2018) Modification of lipase with poly (4-acryloylmorpholine) enhances solubility and transesterification activity in anhydrous ionic liquids 19:1324-1332

Chen X, Liu Z-Q, Lin C-P, Zheng Y-GJBc (2016) Chemoenzymatic synthesis of (S)-duloxetine using carbonyl reductase from Rhodosporidium toruloides 65:82-89

Chen X et al. (2019) Efficient reductive desymmetrization of bulky 1, 3-cyclodiketones enabled by structure-guided directed evolution of a carbonyl reductase 2:931-941

Choi J-M, Han S-S, Kim H-SJBa (2015) Industrial applications of enzyme biocatalysis: Current status and future aspects 33:1443-1454

Chourasia R, Phukon LC, Singh SP, Rai AK, Sahoo D (2020) Role of enzymatic bioprocesses for the production of functional food and nutraceuticals. In: Biomass, Biofuels, Biochemicals. Elsevier, pp 309-334

Ciaramella AJHocn (2019) Psychopharmacology of chronic pain 165:317-337

Conlon HD et al. (1995) Two-step immobilized enzyme conversion of cephalosporin C to 7-aminocephalosporanic acid 46:510-513

Cornils B, Herrmann WA, Xu J-H, Zanthoff H-W (2020) Catalysis from A to Z: a concise encyclopedia. John Wiley & Sons,

de Carvalho CCJMb (2017) Whole cell biocatalysts: essential workers from Nature to the industry 10:250-263

Devine PN, Howard RM, Kumar R, Thompson MP, Truppo MD, Turner NJJNRC (2018) Extending the application of biocatalysis to meet the challenges of drug development 2:409-421

Duggleby HJ, Tolley SP, Hill CP, Dodson EJ, Dodson G, Moody PCJN (1995) Penicillin acylase has a single-amino-acid catalytic centre 373:264-268

Dumorné K, Córdova DC, Astorga-Eló M, Renganathan P (2017) Extremozymes: a potential source for industrial applications

Engleder M et al. (2017) Structure-based reaction mechanism of oleate hydratase from Elizabethkingia meningoseptica

Fernandes PJEr (2010) Enzymes in food processing: a condensed overview on strategies for better biocatalysts 2010

Ferrandi EE, Monti DJWJoM, Biotechnology (2018) Amine transaminases in chiral amines synthesis: recent advances and challenges 34:1-10

Finnigan W, Hepworth LJ, Flitsch SL, Turner NJJNC (2021) RetroBioCat as a computer-aided synthesis planning tool for biocatalytic reactions and cascades 4:98-104

Finnigan W, Hepworth LJ, Turner NJ, Flitsch S (2020) RetroBioCat: Computer-Aided Synthesis Planning for Biocatalytic Reactions and Cascades

Froestl WJFmc (2011) An historical perspective on GABAergic drugs 3:163-175

García-Junceda E, Lavandera I, Rother D, Schrittwieser JHJJoMCBE (2015) (Chemo) enzymatic cascades—Nature's synthetic strategy transferred to the laboratory 114:1-6

Gardossi L et al. (2010) Guidelines for reporting of biocatalytic reactions 28:171-180

Ghislieri D et al. (2018) Engineering an Enantioselective Amine Oxidase for the Synthesis of Alkaloid Natural Products and Pharmaceutical Building Blocks:37

Ghislieri D et al. (2013) Engineering an enantioselective amine oxidase for the synthesis of pharmaceutical building blocks and alkaloid natural products 135:10863-10869

Gomm A, O'Reilly EJCoicb (2018) Transaminases for chiral amine synthesis 43:106-112

Gong J-S, Lu Z-M, Li H, Shi J-S, Zhou Z-M, Xu Z-H (2012a) Nitrilases in nitrile biocatalysis: recent progress and forthcoming research Microbial Cell Factories 11:142 doi:10.1186/1475-2859-11-142

Gong J-S, Lu Z-M, Li H, Shi J-S, Zhou Z-M, Xu Z-HJMcf (2012b) Nitrilases in nitrile biocatalysis: recent progress and forthcoming research 11:1-18

Gopinath V, Murali A, Dhar KS, Nampoothiri KMJAm, biotechnology (2012) Corynebacterium glutamicum as a potent biocatalyst for the bioconversion of pentose sugars to value-added products 93:95-106

Goswami A, Stewart JD (2016) Introduction, Types of Reactions, and Sources of Biocatalysts. In: Organic Synthesis Using Biocatalysis. Elsevier, pp 1-38

Goswami A, Van Lanen SGJMB (2015) Enzymatic strategies and biocatalysts for amide bond formation: tricks of the trade outside of the ribosome 11:338-353

Guan L-J et al. (2015) A new target region for changing the substrate specificity of amine transaminases 5:1-11

Hardianto D, Royani JI, Safarrida AJJPAM (2016) Cephalosporin C acylase from microbes for one-step enzymatic transformation of cephalosporin C to 7-aminocephalosporanic acid 10:2495-2499

Hibbert EG, Dalby PAJMCF (2005) Directed evolution strategies for improved enzymatic performance 4:1-6

Hill RA, Sutherland AJNPR (2016) Natural Product Reports 33:1239

Hollmann F, Fernandez-Lafuente R (2021) Grand Challenges in Biocatalysis 1 doi:10.3389/fctls.2021.633893

Huang PJJ, Liu JJC (2020) In vitro selection of chemically modified DNAzymes 9:1046

Illanes A, Fajardo AJJoMCBE (2001) Kinetically controlled synthesis of ampicillin with immobilized penicillin acylase in the presence of organic cosolvents 11:587-595

Jäckel C, Hilvert DJCoib (2010) Biocatalysts by evolution 21:753-759

Kelly SA, Pohle S, Wharry S, Mix S, Allen CC, Moody TS, Gilmore BFJCr (2018) Application of ω-transaminases in the pharmaceutical industry 118:349-367

Khobragade TP, Sarak S, Pagar AD, Jeon H, Giri P, Yun H (2021) Synthesis of Sitagliptin Intermediate by a Multi-Enzymatic Cascade System Using Lipase and Transaminase With Benzylamine as an Amino Donor 9 doi:10.3389/fbioe.2021.757062

Lan Y et al. (2017) Overexpression and characterization of two types of nitrile hydratases from Rhodococcus rhodochrous J1 12:e0179833

Land H, Humble MS (2018) YASARA: a tool to obtain structural guidance in biocatalytic investigations. In: Protein Engineering. Springer, pp 43-67

Li T et al. (2012) Efficient, chemoenzymatic process for manufacture of the boceprevir bicyclic [3.1. 0] proline intermediate based on amine oxidase-catalyzed desymmetrization 134:6467-6472

Li Z et al. (2011) Pregabalin is a potent and selective ligand for α2δ-1 and α2δ-2 calcium channel subunits 667:80-90

Lin B, Tao YJMCF (2017) Whole-cell biocatalysts by design 16:1-12

Liu Q, Xun G, Feng YJBa (2019) The state-of-the-art strategies of protein engineering for enzyme stabilization 37:530-537

Lorenzen J, Driller R, Waldow A, Qoura F, Loll B, Brück TJC (2018) Rhodococcus erythropolis Oleate hydratase: a new member in the Oleate hydratase family tree—biochemical and structural studies 10:407-414

Luo Y, Chen Y, Ma H, Tian Z, Zhang Y, Zhang J (2017) Enhancing the biocatalytic manufacture of the key intermediate of atorvastatin by focused directed evolution of halohydrin dehalogenase Scientific Reports 7:42064 doi:10.1038/srep42064

Marshall WF, Blair JE The cephalosporins. In: Mayo Clinic Proceedings, 1999. vol 2. Elsevier, pp 187-195

Mazurenko S, Prokop Z, Damborsky JJAC (2019) Machine learning in enzyme engineering 10:1210-1223

McAuliffe JC, Aehle W, Whited GM, Ward DE (2007) Industrial Enzymes and Biocatalysis. In: Kent JA (ed) Kent and Riegel's Handbook of Industrial Chemistry and Biotechnology. Springer US, Boston, MA, pp 1375-1420. doi:10.1007/978-0-387-27843-8_31

Micklitsch C, Duan D, Borra-Garske MJIEA (2019) KREDs: Toward Green, Cost-Effective, and Efficient Chiral Alcohol Generation:351-383

Mohan SV, Modestra JA, Amulya K, Butti SK, Velvizhi GJTiB (2016) A circular bioeconomy with biobased products from CO2 sequestration 34:506-519

Mokgoro MM (2003) Dicarboxylic acid production by Yarrowia lipolytica strains. University of the Free State

Morrison D, Rothenbroker M, Li YJSM (2018) DNAzymes: selected for applications 2:1700319

Mu R et al. (2020) Application of Enzymes in Regioselective and Stereoselective Organic Reactions 10:832

Nagasawa T, Nakamura T, Yamada H (1990) Production of acrylic acid and methacrylic acid using Rhodococcus rhodochrous J1 nitrilase Applied Microbiology and Biotechnology 34:322-324 doi:10.1007/BF00170051

Nagasawa T, Yamada HJTiB (1989) Microbial transformations of nitriles 7:153-158

Novick SJ et al. (2021) Engineering an Amine Transaminase for the Efficient Production of a Chiral Sacubitril Precursor 11:3762-3770

Park H, Park S, Yang Y-H, Choi K-YJCRiB (2021) Microbial synthesis of violacein pigment and its potential applications:1-23

Patel RN (2000) Stereoselective biocatalysis. CRC Press,

Patel RNJAC (2011) Biocatalysis: Synthesis of key intermediates for development of pharmaceuticals 1:1056-1074

Patel RNJCCR (2008) Synthesis of chiral pharmaceutical intermediates by biocatalysis 252:659-701

Pellis A, Cantone S, Ebert C, Gardossi LJNb (2018) Evolving biocatalysis to meet bioeconomy challenges and opportunities 40:154-169

Philp JJE, Science E (2015) Balancing the bioeconomy: supporting biofuels and bio-based materials in public policy 8:3063-3068

Raj J, Sharma NN, Prasad S, Bhalla TCJJoIM, Biotechnology (2008) Acrylamide synthesis using agar entrapped cells of Rhodococcus rhodochrous PA-34 in a partitioned fed batch reactor 35:35

Rosenthal K, Lütz SJCOiG, Chemistry S (2018) Recent developments and challenges of biocatalytic processes in the pharmaceutical industry 11:58-64

Rozzell JDJB, Chemistry M (1999) Commercial scale biocatalysis: myths and realities 7:2253-2261

Schmeisser C, Steele H, Streit WRJAm, biotechnology (2007) Metagenomics, biotechnology with non-culturable microbes 75:955-962

Schoemaker HE, Mink D, Wubbolts MGJS (2003) Dispelling the myths--biocatalysis in industrial synthesis 299:1694-1697

Shahzad D, Faisal M, Saeed A (2019) Chemoenzymatic Approaches towards (S)-Duloxetine. In: Pharmaceutical Biocatalysis. Jenny Stanford Publishing, pp 143-169

Shen J-D, Cai X, Liu Z-Q, Zheng Y-G (2021) Nitrilase: a promising biocatalyst in industrial applications for green chemistry Critical Reviews in Biotechnology 41:72-93 doi:10.1080/07388551.2020.1827367

Stephanopoulos G (1996) Biotechnology, Bioprocessing vol 3. Wiley-Blackwell,

Survase SA, Bajaj IB, Singhal RSJFT, Biotechnology (2006) Biotechnological production of vitamins 44:381-396

Swainston N et al. (2018) STRENDA DB: enabling the validation and sharing of enzyme kinetics data 285:2193-2204

Tao JA, Lin G-Q, Liese A (2009) Biocatalysis for the pharmaceutical industry: discovery, development, and manufacturing. John Wiley & Sons,

Taylor D (2015) The pharmaceutical industry and the future of drug development

Truppo MDJAmcl (2017) Biocatalysis in the pharmaceutical industry: the need for speed 8:476-480

Van Den Burg BJCoim (2003) Extremophiles as a source for novel enzymes 6:213-218

Van Pelt S et al. (2011) Pseudomonas stutzeri lipase: a useful biocatalyst for aminolysis reactions 13:1791-1798

Van Roon J, Schroën C, Tramper J, Beeftink HJBa (2007) Biocatalysts: Measurement, modelling and design of heterogeneity 25:137-147

VIETHT W Continuous Enzymatic Production of Invert Sugar

Von der Osten C et al. (1993) Protein engineering of subtilisins to improve stability in detergent formulations 28:55-68

Wang X, Su R, Chen K, Xu S, Feng J, Ouyang PJFim (2019a) Engineering a microbial consortium based whole-cell system for efficient production of glutarate from L-lysine 10:341

Wang Z, Wu X, Li Z, Huang Z, Chen FJO, chemistry b (2019b) Ketoreductase catalyzed stereoselective bioreduction of α-nitro ketones 17:3575-3580

Willies SC, Galman JL, Slabu I, Turner NJJPTotRSAM, Physical, Sciences E (2016) A stereospecific solid-phase screening assay for colonies expressing both (R)-and (S)-selective ω-aminotransferases 374:20150084

Willis NJ, Fisher CA, Alder CM, Harsanyi A, Shukla L, Adams JP, Sandford GJGC (2016) Sustainable synthesis of enantiopure fluorolactam derivatives by a selective direct fluorination–amidase strategy 18:1313-1318

Wiltschi B et al. (2020) Enzymes revolutionize the bioproduction of value-added compounds: From enzyme discovery to special applications 40:107520

Wu S, Snajdrova R, Moore JC, Baldenius K, Bornscheuer UTJACIE (2021) Biocatalysis: enzymatic synthesis for industrial applications 60:88-119

Xu F, Chen S, Xu G, Wu J, Yang L (2015) Discovery and expression of a Pseudomonas sp. esterase as a novel biocatalyst for the efficient biosynthesis of a chiral intermediate of pregabalin Biotechnology and Bioprocess Engineering 20:473-487 doi:10.1007/s12257-015-0069-1

Yamada H, Kobayashi MJB, biotechnology,, biochemistry (1996) Nitrile hydratase and its application to industrial production of acrylamide 60:1391-1400

Yuryev R, Liese AJC (2010) Biocatalysis: the outcast 2:103-107

Zheng R-C, Wang T-Z, Fu D-J, Li A-P, Li X-J, Zheng Y-G (2013) Biocatalytic synthesis of chiral intermediate of pregabalin with high substrate loading by a newly isolated Morgarella morganii ZJB-09203 Applied Microbiology and Biotechnology 97:4839-4847 doi:10.1007/s00253-013-4810-4

Žnidaršič-Plazl PJCOiG, Chemistry S (2021) Biocatalytic process intensification via efficient biocatalyst immobilization, miniaturization, and process integration 32:100546

6

Immobilized Enzymes

A. Shobana, M. Priyadarshini, R. Shruthi, V. Aishwarya M. Rajeswari, A. Sujatha and Anitha Subash

Department of Biochemistry, Biotechnology and Bioinformatics Avinashilingam Institute for Home Science and Higher Education for Women Coimbatore, Tamil Nadu, India

Abstract

Enzyme is a highly efficient catalyst that is produced by active cells. When compared to chemical catalysts; enzymes have a number of benefits, including high specificity, high catalytic efficiency and flexible activity, all of which greatly help to promote their use in the pharmaceutical, chemical, and food sectors. The majority of enzymes such as protease, amylase, lipase and cellulase are relatively unstable and industrial uses are frequently restricted by a lack of long-term operational stability as well as the technically challenging recovery process and ruse of the enzyme. Enzyme immobilization provides an excellent base for increasing enzyme availability to the substrate with higher turnover over a long period of time. Enzyme immobilization are done by several methods and their performance are influenced by many factors. Physical and chemical based immobilization are the most commonly used methods. The matrix immobilizes the enzyme by holding it permanently or temporarily for a short stretch of time. The generally used matrixes for immobilizations includes natural, synthetic polymers and inorganic compounds.Currently, synergistic interactions of biotechnology and nanotechnology are useful for the development of immobilized enzymes. Enzyme immobilization is a commonly used technology in a variety of industries including food, pharmaceuticals, bioremediation, detergents, and textiles due to technical and economical advantages. This chapter focuses on most commonly used immobilized techniques in industry with many recent applications of immobilized enzymes.

Introduction

Enzymes are proteins that aid in the speeding up of a variety of biochemical and chemical reactions. They are natural catalysts and found in plants, animals, and microorganisms, where they accelerate essential biological processes (Homaei et al., 2013).

In recent years, the applications of enzymes in various industries has been steadily rising. Enzyme applications in industries includes food (baking, dairy products, starch conversion), textiles, pulp and paper, detergents, biosensors, cosmetics, wastewater treatment, animal feed, pharmaceuticals and chemical manufacture and more recently production of biofuels such as biodiesel and bio-ethanol (Homaei, 2016). Different approaches for cost reduction have been implemented in an attempt to utilize the enzymes in biotechnological processes. One of the most favorable methods is immobilization (Mohamad et al., 2015). Immobilized enzymes have been utilized from 1916, when Nelson and Griffin found that invertase when absorbed to charcoal undergo hydrolysis of sucrose. Grubhofer and Schelth revealed the possibility of immobilized enzymes for reuse and stability and reported the covalent immobilization of many enzymes (Hemalatha et al., 2016).

Enzyme immobilization entails combining the enzyme's selectivity, stability and kinetics with the physical carriers and chemical properties in unique formulations with the primary goal of increasing the biocatalyst's physical and enzymatic stability (Basso and Serban, 2019). The immobilization of an enzyme can protect it from the external environment and structural denaturation. The immobilized enzymes can be simply detached from the reaction solution after assessing the activity. The immobilized enzymes can be washed and reused (Lee et al., 2020). The advantages of enzyme immobilization include continuous economic operations, automation, a high investment-to-capacity ratio and high purity product recovery (Datta et al., 2013).

Different types of methods such as adsorption, entrapment, ionic bonding, and covalent binding are used to immobilize enzymes (Hassan et al., 2019). In reaction conditions, all enzyme immobilization carriers must be non-toxic, biocompatible, and insoluble. It should possess high affinity for protein, contains appropriate functional groups and thermal stability (Yushkova et al., 2019). The immobilized enzymes maintain high catalytic activity by protecting the enzyme structure from harsh reaction conditions (Zdarta et al., 2018). This chapter focuses on most commonly used immobilized techniques in industry with many recent applications of immobilized enzymes.

Modes of Immobilization

Enzyme immobilization can be defined as the incarceration of the enzymes onto a matrix either physically or chemically by which they can retain their full activity or most of its activity. The immobilization of enzymes allows the enzymes to retain their activity even at various temperatures and pH and makes them stationary throughout the processes and also protects the enzymes from various environmental factors and impact of the enzyme catalyzed reaction (Imam et al., 2021). There are different types of immobilization. They are:

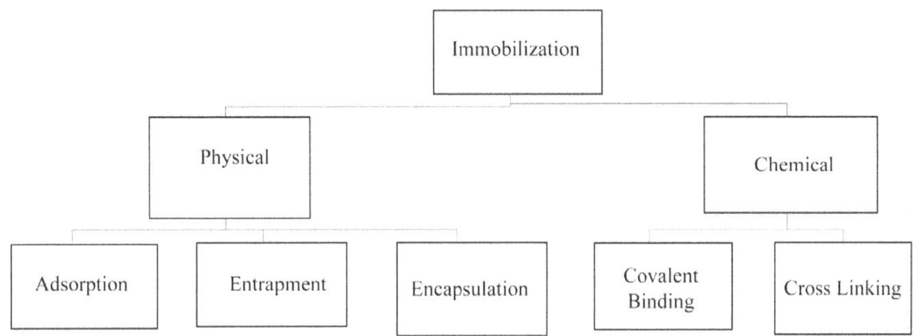

Fig. 1: Modes of immobilization

Adsorption

Physical adsorption is one of the reversible immobilization techniques where the enzyme gets physically attached to a solid support or a matrix. The adsorption of enzymes could occur by weak nonspecific forces such as hydrogen bonds, van der waals force and hydrophobic interactions. By this method, solid support could be retained after the use and could be filled with another enzyme for use. The carrier molecule will be organic or inorganic.

Entrapment

Physical entrapment is the method where the enzyme is entrapped into a solid matrix or 3-D matrix of high molecular compounds. The bonds that are involved in the entrapment of enzymes are covalent or non-covalent bonds and the matrix used is a water soluble polymer. There are two different types of entrapment. They are:

1. Gel entrapment
2. Microcapsule entrapment

Gel Entrapment

Commonly used gels for entrapment are calcium alginate, polyacrylamide gel, gelatin etc. It is the entrapment of an enzyme in the gel lattice where it does not require any modifications of the amino acid residue (Ann-Christin and Koch-Schmidt, 1977).

Microcapsule Entrapment

Microcapsule entrapment is an immobilization method where enzymes are encapsulated inside a coating such as gelatin, sodium alginate etc. This technique is mainly used to incorporate food ingredients, enzymes and cells and is mainly used in pharmaceuticals. Sodium alginate is the most commonly used method to immobilize enzymes, wherein it forms beads and entraps the enzyme in it (Aeron and Morya, 2017).

Encapsulation

In the encapsulation method, enzymes are enclosed in a semi-permeable membrane like the nitrocellulose, which prevents the enzyme leakage or direct contact with the environment. The larger molecules like protein and cells cannot interact with the encapsulated enzymes. Permeable substances diffuse in and out of the membrane by acting upon the enzyme inside it. These encapsulated enzymes will be more stable under various conditions that are used in catalysis, enzyme therapy and biosensors (Bialas et al., 2021).

Covalent Binding

The covalent binding method is one of the most widely used immobilization methods which provides a binding between the enzymes and the support solid matrix and forms a stable complex. The chemical groups that form covalent bonds are phenol ring, amino, thiol, hydroxyl, imidazole and carboxyl groups. Of these, hydroxyl group forms the covalent bond more easily than other groups (Vijayalakshmi et al., 2020).

Cross-Linking

Cross linking is otherwise called as co-polymerization method, which does not require any solid support material, where the enzymesare directly linked by the covalent bonds through the polyfunctional reagents. Glutaraldehyde and diazonium salt arethe most commonly used polyfunctional reagents. It does not involve highly purified enzymes for immobilization (Yamaguchi et al., 2018).

Materials used for Fabrication of Immobilization Supports

Solid support materials are used to immobilize enzymes. The solid support materials are separated into two different types based on their chemical composition: (a) inorganic support material and (b) organic support material. Organic support material can be divided into natural and synthetic matrices (Sirisha et al., 2016).

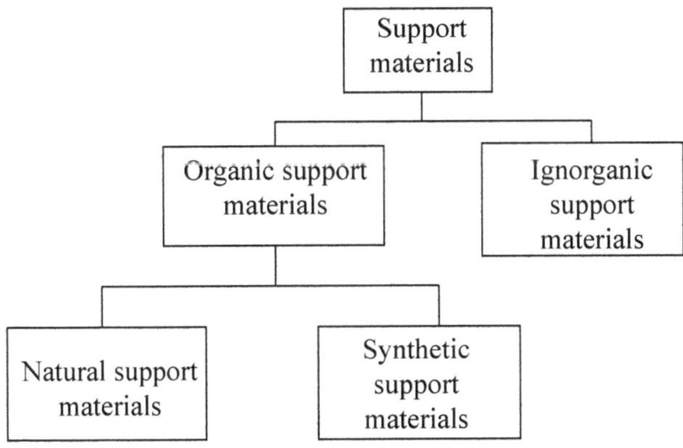

Fig. 2: Types of Immobilization supports

Inorganic Support Materials

Silica and Inorganic Oxides

The most commonly utilized inorganic support material was the silica. Silica matrix provides excellent resistance to heat and chemicals, good mechanical qualities and is the good matrix for various applications. Because of its huge surface area and porous structure, silica has good sorption properties. Silica-gel, colloidal silica nanoparticles, Sol-gel and fumed silica have been used to encapsulate enzymes such as oxidoreductases, transferases, hydrolases, and isomerases (Zdarta et al., 2018).

Ceramics

The word "ceramics" refers to insoluble, inorganic, solid, non-metallic material made from metal, mixed metallic and non-metallic oxides. The ceramics are produced by heating plastic and moldable (water-containing) beginning mixtures. The porosity of ceramics varies substantially depending on the initial material, heating duration and temperature. Heating of ceramics for long periods usually results in a thickening of the materials with concomitant sintering, which greatly reduces the porosity (Zucca and Zanjsut, 2014).

Glass

Glass matrix was found to be an intriguing trapping medium for various applications. Glass is used as plates and beads in physical forms. Furthermore, rather than employing external anchoring agents such as beads or rings, numerous authors reported that the glass was designed to construct the photocatalytic reactor (Goutham et al., 2019).

Activated Carbon

Carbon based materials are considered as an excellent support material for immobilizing enzymes. Enzyme immobilization has been successfully carried out by using activated carbons, graphene and biochars (Adamian et al., 2021). Activated carbon is tasteless, non-toxic, high regeneration, low adsorption energy and long – lasting material. Activated carbon (AC) is mostly used to treat wastewater. The abundant pores present in the activated charcoal have the ability to adsorb contaminants such as micro-pollutant, dyes, organic and inorganic compounds, etc.

Protein Coated Microcrystals Technique

Protein coated microcrystals are one of the inorganic support materials used for immobilization methods. Protein-coated microcrystal technique is widely used for the lyophilized enzymes by the addition of lyoprotectants and other mineral salts. They are prepared by mixing an aqueous solution of an enzyme with any salt such as potassium sulphate, amino acid or sugar and vigorously mixing it and drop into the solvent such as isopropyl alcohol so that microsized crystals are produced (Ali et al., 2017).

New Carriers Used In Immobilization Techniques

Enzymes have also been immobilized using nanoporous gold and nanotubes support materials. The production of immobilized enzymes using nanomaterials were used in the electrode fabrication and biosensor applications (Elnashar, 2010).

Natural Polymers

Alginate

One of the most important methods of immobilization for enzyme entrapment is calcium alginate. Alginates are water-soluble and the properties are thickening, emulsifying, and film-forming agents that have been utilized in the food and pharmaceutical sectors for over 65 years. Entrapment of enzymes and cells in insoluble calcium alginate gels is known as a quick, safe, low-cost and variable approach for immobilizing enzymes and cells (Anwar et al., 2009).

Sodium alginate is a type of natural polymer obtained from plants. Sodium alginate is a low-cost polymer among natural polymers. Sodium alginate can be cross-linked with calcium ions to form sodium alginate gel spheres. Physical crosslinking of sodium alginate and calcium ions are mild and the reaction speed is fast (Bai and Wu, 2021).

Chitin and Chitosan

Chitin and its derivatives-based supports provide stability and cost-effectiveness. Chitosan consists of numerous functional groups and is biocompatible, biodegradable, and non toxic. In comparison to free enzymes, chitosan bound enzymes have improved biocatalytic performance due to their excellent operational stability and reusability (Verma et al., 2019).

Gelatin

Intensive studies were done in gelatin because of its cost effectiveness, biodegradability, biocompatibility and non-immunogenic properties. Thermal and chemical degradation of collagen yields gelatin. It has weak mechanical properties and limits its potential applications as a biomaterial. At the same time, due to the huge number of functional side groups, it is easily crosslinked with chemicals. The crosslinking process improves the mechanical and thermal stability of gelatin materials (Pulat and Akalin, 2012).

Carrageenan

Entrapment methods have been utilized to immobilize enzymes and cells using carrageenan. It is cost effective and has poor mechanical and thermal stability (Elnashar et al., 2014).

Cellulose

Cellulose as a support material confers various advantages, including substantiality, attainability, biodegradability and low cost, all of which promote the production of lightweight and environmental friendly composites (Gabr et al., 2014). Under physiological conditions, cellulose exhibits crucial properties like biocompatibility, amphiphilic and chemically inert (Gennari et al., 2020). Recently, researchers found out various methodologies for the preparation of cellulose-based nanostructures in the last ten years (Jackson et al., 2019).

Synthetic Support Material

Synthetic polymers are produced by means of chemical polymerization using various monomers. Synthetic polymer materials, as a type of key carrier, have the advantages of high mechanical stiffness, specific surface area, ease of changing surface properties and the ability to bring certain functional groups according to actual needs (Zhang et al., 2013).

Recently, the synthetic support materials called as core-shell nanoparticles were produced by P4VP Poly (4-vinylpyridine) to immobilize glucose oxidase and horseradish peroxidase on single particles, resulting in a 20% increase in enzyme activity compared to free enzymes. The second example is polystyrene, which is one of the most inexpensive synthetic polymers and is currently available in a variety of sizes. Gigaporous and macroporous polystyrene microspheres are types of synthetic materials to immobilize the enzyme lipase. The reusability is impressive and even after 100 recycling, the activities can remain nearly 100% and 93% (Lyu et al., 2017).

Applications of Immobilized Enzymes

Detergent Industry

The enzymes based detergents have notable cleaning properties other than the available synthetic detergents. These detergents are eco- friendly and are active at cold temperature (Kumar et al., 1998). These enzymes are required in small quantities for washing and the activity is not lost after the stain removal and also found to have improved quality of the fabric.

Table 1: General enzymes used in the detergent industry

S.No	Enzymes	Applications
1.	Protease	In detergent industry to remove protein stains such as grass, blood, egg and human sweat which have a tendency to adhere strongly to textile fibers.
2.	Amylases	To remove residues of starch-based foods like potatoes, spaghetti, custards, gravies and chocolates.
3.	Lipase	To decompose fatty material. To remove fat related stains such as fats, butter, salad oil, sauces and the tough stains on collars and cuffs.
4.	Cellulase	In textile industry- to modify the structure of cellulose fiber on cotton and cotton blends. In detergent industry- color brightening, softening and soil removal.

Source: Hasan et al., 2006

When non-enzymatic detergents are used stains become permanent because of oxidation and denaturation process caused by bleaching and drying. When it is combined with surfactants and enzymes hard stains can be removed easily. Extracellular enzyme produced by alkaliphilic *Bacillus* strains is used in the detergent industry. Other enzymes like proteases, alpha amylases and cellulases are produced on a large scale and have been used in detergents (Ito et al., 1998).

Food Industry

Immobilized enzymes have extensive range of industrial uses particularly, food industry. Immobilized enzymes are used to tenderise meat and meat products, to make protein hydrolysate, to clarify juice and beer in the brewing sector, to make cheese in the dairy sector, to make bread, and to extract flavour and colour compounds from plants (Homaei, 2015).

Dairy Industry

Lactose-free milk and milk products are more acceptable to lactose-intolerant people, hence the dairy industry has invested much in developing beta-galactosidase to hydrolyze lactose. Lactose is converted to glucose and galactose by the action of the enzyme galactosidase (EC 3.2.1.23), also known as lactase. It is also used to eliminate lactose from milk (Basso and Serban, 2019). The immobilization of galactosidase results in the hydrolysis of lactose in the dairy sector has piqued researcher's interest. Although most companies still employ free enzymes to hydrolyze lactose, immobilization of β-galactosidase is a challenging issue due to its fringe benefit (Panesar et al., 2010).

Brewing Industry

Beverage industry requires 6–7 days of fermentation period for the production of beers. In the brewing industry, the immobilization technique involves encasing yeast cells in order to improve their concentration. The use of immobilized yeast cells shortens the processing time without reducing the product quality (Adhikari, 2021).

Cocoa Industry

Cocoa butter is a valuable substance with desirable properties including melting point, snap, and gloss. Transesterification reactions use lipases to transform palm oil into cocoa butter and can be used to modify cost effective fats like shea butter, salt fat, and palm oil, resulting in low-cost alternatives for cocoa butter (Mandake et al., 2020).

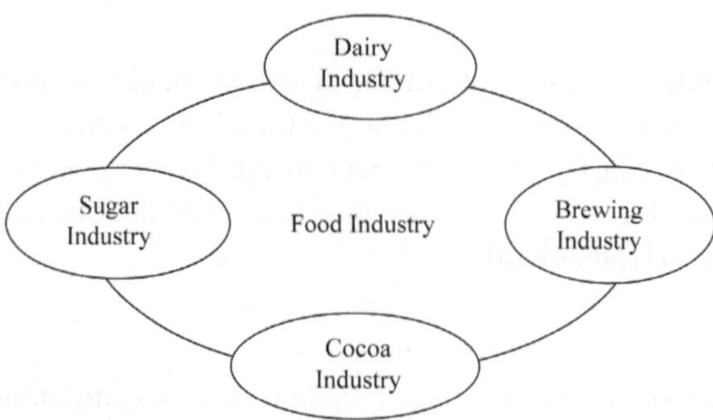

Fig. 3: Applications of immobilized enzymes in Food industry

Immobilized Enzymes in Other Food Industry

The enzymes that are immobilized from cells were used for the production of alcohol. Bacterial cells that produce alcohol are immobilized, and fermentation takes place in a bioreactor (Krishnamoorthi et al., 2015).

The transformation of glucose to fructose results in the yield of high fructose syrup and it is one of the most notable industrial processes that uses immobilized enzymes. Immobilized glucose isomerase is currently being utilized in making glucose high fructose syrup, which comprises 42% fructose, 50% glucose and 8% additional sugars. In the upcoming years, an amyloglucosidase-based immobilization technique may be used to address the rising demand for glucose as a substrate for the yield of high fructose syrup (Aggarwal and Sahini, 2012).

Textile Industry

A range of enzymes are used in the textile industries. Protease, amylase, cellulase, lipase, catalase and laccase are the industrial enzymes mainly used in textile processes. These enzymes act mainly as the stain removers and reduce the usage of hydrogen peroxide, lignin and bleaching textiles. They are also used in processes such as de-sizing, scouring, bleaching, dyeing and finishing. Since textile industries need highly stable enzymes and should perform very good at extreme temperature and pH, many strategies are evolving to immobilize the enzymes and extend their shelf life for the performance (Soares et al., 2011).

Pharmaceutical Industry

The production or processing of enzymes for use as drugs is an important secondary part of today's pharmaceutical industry. Every research centre of pharmacy in the world made attempts to use the benefits of enzymes as medicines. Immobilized enzymes are generally used in the medical field to diagnose and treat diseases (Hassan et al., 2019). Immobilized enzymes can also be used to treat some disorders, such as inborn metabolic abnormalities, cardiovascular diseases, cancer, intestinal illnesses as well as intoxication (Pastor et al., 2013). Enzyme encapsulation via electroporation is the most simple and effective method of immobilization in the biomedical area and it is a reversible procedure that allows the enzyme to be regenerated. (Hemalatha et al., 2016).

Biosensors

Immobilized enzyme biosensors are a form of catalytic biosensor in which enzymes are immobilized on the transducer surface and act as a bridge between the transducer and the analyte. Biosensors based on immobilized enzymes have been widely used in a variety of sectors, including biomedical applications (Nguyen et al., 2019).

The immobilization of glucose oxidase (GOx) is critical for the construction of electrochemical biosensors. Electrochemical sensors based on GOx immobilization remain the most common commercial approach to glucose monitoring in diabetes treatment and progression. Immobilization of GOx on nanomaterials for sensor development is a growing field, therefore immobilization on nanoparticles-Au is being investigated (Stine, 2017).

Table 2: Some important medical uses of immobilized enzymes

S.No	Enzymes	Medical uses
1.	Arginase	Cancer
2.	Asparaginase	Leukemia
3.	Catalase	Treat mutations in catalase gene
4.	Glucoamylase	Glycogen storage disease
5.	Glucose oxidase	Artificial pancreas
6.	Urease	Artificial kidney
7.	Xanthine oxidase	Lesch–Nyhan disease

Source: Homaei et al., 2013

Wastewater Treatment

In today's environment the most important pollutant is the dyes that are released from the industries. They are released directly into the lakes, sea and ponds without proper treatment of and makes the water unfit for further use. The dyes that are toxic to the environment are anthraquinone, azo and triarylmethane, Direct blue 15, Disperse blue 291 which can affect the DNA of the organism. Thus, it is important to produce immobilized enzymes which is effective in removing dyes from wastewater (Zdarta et al., 2021)

Other Industries

Immobilized enzymes are also used in other industries like flavors and aroma industry, bioremediation, and also for the production of biodiesel.

Flavors and Aromas Industry

Biocatalytic techniques are being developed for larger production of fragrance, flavour and aromatic compounds that are more cost-effective and eco- friendly than traditional chemical processes. The immobilized lipase plays a key role in the development of biocatalytic fragrance and flavour generation because of their ability to transfer acyl groups from esters to other nucleophiles (Chapman et al., 2018).

Bioremediation

Immobilized oxidative enzymes are widely considered as a green solution to the problem of high levels of micropollutants in the environment. When compared to free enzymes, immobilizedlaccase or horseradish peroxidase exhibited high stability and reusability, as well as facile separation from reaction mixture, making them more attractive and cost-effective. The immobilized laccase or horseradish peroxidase has the ability to oxidize compounds (Shakerianet al., 2020).

Biodiesel

The principal options for large-scale application of enzymatic biofuel generation are two types of enzymes: cellulases and lipases. Enzyme-based hydrolysis can be made more cost-effective by increasing the thermal stability and reusability of enzymes, which can all be assisted by immobilizing enzymes on support matrices. The use of silica and polymeric nanoparticles to immobilize cellulases for use in biofuel production has been investigated (Puri et al., 2013). Immobilized lipase LS-10A isolated from *Candida* sp. from Beijing CTA New Century Biotechnology Co., Ltd has been widely commercialized for biodiesel production (Zhang et al., 2013).

Advantages of Immobilized Enzymes

1) Reusability
2) Continuous use
3) Increased functional efficiency
4) Enhanced reproducibility

5) Needs less labour input
6) Requires minimum reaction time
7) More stable
8) Has high enzyme to substrate ratio
9) Improved process control

Disadvantages of Immobilized Enzymes

1) High cost
2) Limited industrial applications
3) Reduced catalytic properties
4) Some enzymes become unstable after immobilization
5) Gets deactivated in high temperature systems

Conclusion

In recent years,to reduce the production cost and to develop novel industrial processes based on biotransformation enzyme immobilization is developing as a powerful tool. Strategies that are carrier bound and carrier-free offer novel alternatives for intensive and extensive enzyme use at a large scale (Cacicedo et al., 2019). Enzyme stabilization is frequently improved by immobilization on surfaces (i.e. reactor support material), which also allow for re-use of the enzyme, but these immobilization strategies have to be combined with co-factor immobilization, stabilization, and regeneration strategies for industrial relevance (Minteer, 2016). Currently, immobilized enzymes are favored over the free enzymes owing to their extended availability that limits redundant downstream and purification processes (Bhatia, 2018).

References

Adamian, O., Lonappan, L., Alokpa, K., Agathos, S.N. and Cabana, H. (2021), Recent Developments in the Immobilization of Laccase on Carbonaceous Supports for Environmental Applications - A Critical Review, Frontiers in Bioengineering and Biotechnology, 9: 1-29.

Aggarwal, S. and Sahni, S. (2012), The Commercial Exploitation of Immobilized Enzymes, International Conference on Environmental, Biomedical and Biotechnology, 41: 18-22.

Anwar, A., Qader, S.A.U., Raiz, A., Iqbal, S. and Azhar, A. (2010). Calcium Alginate: A Support Material for Immobilization of Proteases from Newly Isolated Strain of Bacillus subtilis KIBGE-HAS, World Applied Sciences Journal, 7 (10): 1281-1286.

Bai, Y.and Wu, W. (2021), The Neutral Protease Immobilization: Physical Characterization of Sodium Alginate-chitosan Gel Beads, Research Square.

Basso, A. and Serban, S. (2019). Industrial applications of immobilized enzymes—A review, Molecular catalysis, 479: 1-20.

Cacicedo, M.L., Manzo, R.M., Municoy, S., Bonazza, H.L., Islan, G.A., Desimone, M., Bellino, M., Mammarella, E.J. and Castro, G.R. (2019), Immobilized Enzymes and their Applications, Advances in Enzyme Technology , 169-200.

Chapman, J., Ismail, A.E. and Dinu, C.Z. (2018), Industrial Applications of Enzyme A reviews: Recent Advances, Techniques, and Outlooks, Catalysts, 8(238): 1-2.

Datta, S., Christena, L.R. and Rajaram, Y.R.S. (2013), Enzyme immobilization: an overview on techniques and support materials, Biotech, 3:1–9.

Elnasha, M.M., Awad, G.E., Hassan, M.E., Eldin, M.S.M., Haroun, B.M. and El-Diwany, A.I. (2014), Optimal Immobilization of β-Galactosidase onto κ-Carrageenan Gel Beads Using Response Surface Methodology and Its Applications, The Scientific World Journal, 1-7.

Elnashar, M. (2010), Review Article: Immobilized Molecules Using Biomaterials and Nanobiotechnology, Journal of Biomaterials and Nanobiotechnology, 1, 61-76.

FarihaHasan, Aamer Ali Shah, SundusJaved and Abdul Hameed, (2010), AfricanJournal of Biotechnology Vol. 9(31), pp. 4836-4844.

Gauri Aeron, Shiwangi M (2017) Immobilization and microencapsulation. Journal of Advance Research in Biotechnology 2 (3): 1-4.

Goutham, R., Narayan, R.B., Srikanth, B. and Gopinath, K.P. (2019), Supporting Materials for Immobilization of Nano-photocatalysts, anophotocatalysis and Environmental Applications. Environmental Chemistry for a Sustainable World, 29: 49-82

Hasan F, Shah AA, Hameed A (2006). Industrial applications of microbial lipases. Enzyme Microbial. Technol. 39: 235-251.

Hassan, M.E., Yang, Q. and Xiao, Z. (2019), Covalent immobilization of glucoamylase enzyme onto chemically activated surface of κ-carrageenan, Bulletin of the National Research Centre, 43(102): 1-11.

Hassan,M.E.,Yang, Q., Xiao, Z.,Liu,L., Wang, N., Cui,X. and Yang, L. (2019), Impact of immobilization technology in industrial and pharmaceutical applications, 3 Biotech, 9(12): 440.

Hemalatha, V., Kalyani, P, Vineela.K, C. and Hemalatha, K.P.J. (2016). Methods, Applications of Immobilized enzymes- A Mini Review, International Journal of Engineering Sciences & Research Technology, 5(11): 523-526.

Homaei, A. (2016). Enzyme Immobilization and its Application in the Food Industry, Advances in Food Biotechnology, 145 - 163.

Homaei, A.A., Sariri, R., Vianello, F. and Stevanato, R. (2013). Enzyme immobilization: an update, National Library of Medicine, 6(4):185-205.

Jackson, E., Correa, S.D. and Betancor, L.(2019), Cellulose-Based Nanosupports for Enzyme Immobilization, Cellulose-Based Superabsorbent Hydrogels, 1235-1250.

Jacub Zdarta et al., (2021), Enhanced Wastewater Treatment by Immobilized Enzymes, current pollution reports, 167-179.

Krishnamoorthi, S., Banerjee, A. and Roychoudhur, A. (2015).Immobilized enzyme technology: potentiality and prospects, Journal of Enzymology and Metabolism, 1(1):104.

Kumar CG, Malik RK, Tiwari MP (1998). Novel enzyme-based detergents: An Indianperspective. Curr. Sci. 75: 1312-1318.

Lee, C.H., Jin, E.S., Lee , J.H. and Hwang, E.T. (2020). Immobilization and Stabilization of Enzymes in Biomineralized Calcium Carbonate Microspheres, National Library of Medicine, 8: 1-8.

Mandake, M.B., Das, U., Phad, L. and Busamwar, S. (2020). Application of Enzyme Immobilization in the Food Industry, International Journal of Advanced Research in Chemical Science, 7(2): 6-10.

Mohamad, N.R., Marzuki, N.H.C., Buang, N.Z., Huyop, F. and Wahab, R.B. (2015). An overview of technologies for immobilization of enzymes and surface analysis techniques for immobilized enzymes, National Library of Medicine, 29(2):205-220.

Nguyen,H.H., Lee, S.H., Lee,U.J., Fermin, C.D.andKim, M. (2019). Immobilized Enzymes in Biosensor Applications, Materials, 12(1): 121.

Panesar, P.S., Kumari,S. andPanesar, R. (2010). Potential Applications of Immobilized-Galactosidase in Food Processing Industries, Enzyme Research, 1-16.

Pulat, M. and Akalin, G.O. (2012), Preparation and characterization of gelatin hydrogel support for immobilization of Candida Rugosa lipase, Artificial Cells, Nanomedicine, and Biotechnology, 42(3): 145 -151.

Puri, M., Barrow, C.J. and Verma, M.L. (2013). Enzyme immobilization on nanomaterials for biofuel production, Trends in Biotechnology,31(4), 1-2.

Shakerian, F., Zhao, J. and Li, S. (2020). Recent development in the application of immobilized oxidative enzymes for bioremediation of hazardous micropollutants - A review, Chemosphere, 1-12.

Sirisha, V.L., Jain, A. and Jain, A. (2016). Enzyme Immobilization: An Overview on Methods, Support Material, and Applications of Immobilized Enzymes, Advances in Food and Nutrition Research, 2017.

Soares, J.C., Moreira, P.R., Queiroga, A.C., Morgado, J., F. Xavier Malcata, F.X. and Manuela E. Pintado, M.E. (2011). Application of immobilized enzyme technologies for the textile industry: a review, Biocatalysis and Biotransformation, 29 (6): 223-237.

Stine, K.J. (2017). Enzyme Immobilization on Nanoporous Gold: A Review, A Biochemcial Insights, 10: 1–12

Verma, M.L., Kumar, S., Das, A., Randhawa, J.S. Chamundeeswari. M. (2019). Chitin and chitosanbased support materials for enzyme Immobilization and biotechnological applications, Environmental Chemistry Letters

Yushkova, E.D., Nazarova, E.A., Matyuhina, A.V., Noskova, A.O., Shavronskaya, D.O., Vinogradov, V.V., Skvortsova, N.N. and Krivoshapkina, E. (2019), Application of Immobilized Enzymes in Food Industry, Journal of Agricultural and Food Chemistry, 5-65.

Zdarta, A., Meyer, A.S., Jesionowski, T. and Pinelo, M. (2018), A General Overview of Support Materials for Enzyme Immobilization: Characteristics, Properties, Practical Utility, Catalysts, 8, 92: 1-27.

Zhang, D.H., Yuwen, L. and Peng, L. (2013).Parameters affecting the performance of Immobilized enzymes, Journal of Chemistry, 1 - 7.

Zucca, P. andSanjust, E. (2014), Inorganic Materials as Supports for Covalent Enzyme Immobilization: Methods and Mechanisms, Molecules, 19(9): 14139–14194.

7

Prospects of Fermentation Technology

Urvashi Srivastava, Unaiza Iqbal, Pinki Saini and Vinita Puranik

Centre of Food Technology, University of Allahabad, Prayagraj Uttar Pradesh

Abstract

Fermentation technology holds immense promise in diverse fields. It offers opportunities for sustainable biofuel production, allowing us to reduce our reliance on fossil fuels and promote a greener energy sector. In the pharmaceutical industry, fermentation plays a vital role in manufacturing antibiotics, enzymes, vaccines, and therapeutic proteins, paving the way for more efficient and cost-effective production processes. Furthermore, fermentation can be harnessed for waste management and environmental remediation, enabling the transformation of organic waste into valuable biogas. With ongoing advancements, fermentation technology continues to open new avenues for innovation and contribute to a more sustainable and environmentally conscious future. The present chapter compares types of fermentation process as well as gives an insight into various industrial prospects of fermentation technology.

Introduction

The onus and imperative for continued development of appropriate technologies for food production have continually risen over the past few years. While conventional food processing techniques still play an important role in the formulation of traditional diets, increasing consumer demand for high-quality, nutritious and safe products propels the industry to search for improved processes. Fermentation remains an age-long food processing technology, practiced, even before the understanding of the underlying processes involved. The techniques and associated knowledge involved in this process are normally handed down from one generation to the other and subsequently passed on within the local communities. The word fermentation is derived from the Latin verb fevere which means "to boil" and fermentation was described by Louis Pasteur as "La vie sans l'air" (life without air). Fermentation technology is defined as a field that involves the use of microbial enzymes in the production of compounds that work within the production of energy, materials, pharmaceutical industries, chemicals, and food industries (Singh et al., 2017). It is a biological process in which various bacteria such as yeast, bacteria, and fungi are involved in the conversion of a complex substrate into a simple, useful human compound (such as metabolites, biomass, recombination technology, and biotransformation) product on industrial scale. People have used it since ancient times for preservation and enhancing organoleptic properties of food. It is a well-established ancient technology used for food preservation, bread production, beer, vinegar, yogurt, cheese, and wine. Over time, it has been refined and varied (Motarjemi 2002). Organic acid and alcohol are the main products of fermentation. In this process, there is the release of secondary metabolites such as antibiotics, enzymes, and growth factors (Subramaniyam et al., 2012; Machado et al., 2004). Various bioactive compounds contain secondary metabolites, for example phenolic compounds, growth factors, antibiotics, mycotoxins, and alkaloids (Martins et al., 2011; Nigam et al., 2009). Recently, there has been an increased demand for fermented foods as potential sources of functional foods. The need to meet consumer demand has made it essential to improve conventional fermentation methods with advanced techniques to ensure the delivery of desired fermented foods with consistently better quality, sensory attributes and nutritional benefits.

1) $C_6H_{12}O_6 \longrightarrow 2CO_2 + 2\ C_2H_5OH$
Alcoholic Fermentation by Yeasts

2a) $2C_2H_5OH \longrightarrow 2CH_3CHO + 2H_2$
Oxidation (Anaerobic) by AAB

2b) $2CH_3CHO + O_2 \longrightarrow 2CH_3COOH + 2H_2O$
Oxidation (Aerobic) by AAB

Eq 1. Showing the steps involved in fermentation

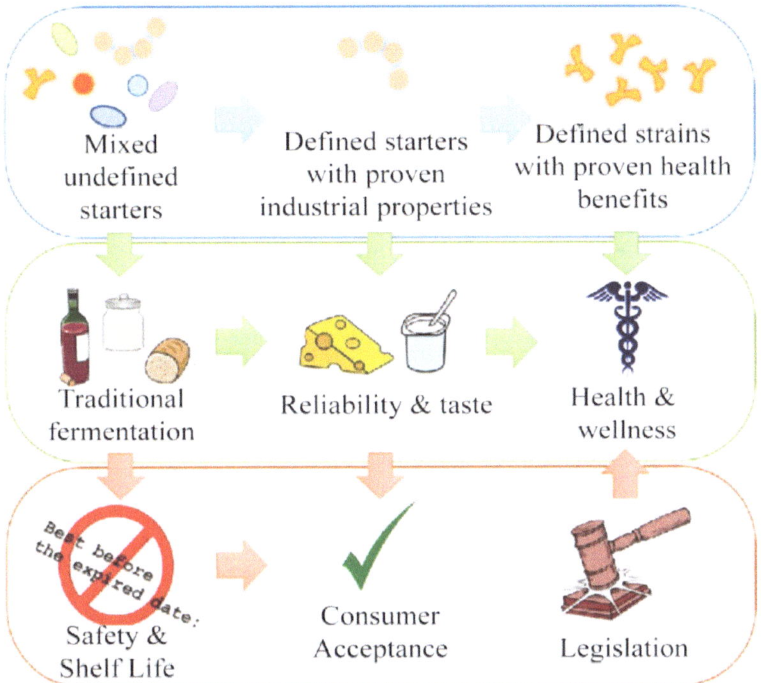

Fig 1: Schematic diagram representing the relationships among fermenting microbes, fermented products, and the consumer.

History of Fermentation

Fermentation is a process that produces energy in which a combination of organisms (usually pyruvate or pyruvate intermediates) acts as both an electron donor and a lethal electron receiver. The Latin root word "fervere" means "to boil" and indicates the release of carbon dioxide when the yeast breaksdown fruit sugar. This process was used to produce alcohol long before humans realized that it was actually a fermentation process.

The people of ancient Egypt were the first to boil alcohol and vinegar in malt grains and fruit juices during the 1700s. However, they had no understanding of the contamination process and the role of microorganisms in fermentation, and the use of certain production equipment had not yet been discovered. In the middle of the eighteenth century, Louis Pasteur established the role of yeast in the fermentation of wine, and from then on the pure tradition was used for high-quality wine, beer, and vinegar. At the same time, Carlsberg tried to separate the yeast into cells to produce high quality alcohol. In the late eighteenth and early nineteenth centuries, the use of aerator in vinegar production began. The fermentation of the vinegar included a well-known amount of non-pasteurized vinegar, pre-fermented as the first component to prevent contamination and to act as a stove. Thus, control of the fermentation process was initiated. The importance of fermentation and production of various products began to be considered in the early nineteenth century and continued into the mid-nineteenth century. Several products are produced by fermentation, namely, glycerol, baker's yeast, lactic acid, and acetone butanol. The production of secondary metabolites by fermentation (such as antimicrobials, amino acids, and vitamins) flourished in the late nineteenth century due to the need to treat soldiers who fought in World War II. Submerged underwater fermentation under aerobic conditions with moderate process control was established during this period. In subsequent years, the fermentation industry has seen continuous jumps and bumps in the production of high-value metabolites, including various antibiotics and growth hormones, using complex bioreactors (Humer and Schedule, 2016). Current fermentation technology focuses on in vitro production of animal cells, monoclonal antibodies, and insulin. The sector is developing rapidly using technologies from different sectors.

Types of Fermentation

Fermentation is one of the oldest scientific domains which has evolved, refined and been diversified over the centuries. Fermentation processes are mainly classified as either (A) solid state or (B) submerged culture. Solid state fermentation (SSF) differs from submerged fermentation (SmF) with respect to flowing water, which is present in SmF while it is absent or very minor in SSF. In the following sections, we have discussed various aspects of solid-state and submerged fermentation processes.

Solid State Fermentation (SSF)

The growth of microorganisms on moist solid substrate particles in the absence or minimum water between the particles is known as solid state fermentation. The moisture content of solid substrate ranges between 12–80%. SSFs are usually used for the fermentation of agricultural products or foods. During SSF, the substrate may require preparation or pretreatment, like chopping or grinding-reduce particle size, cooking or chemical hydrolysis, pasteurization or sterilization so as to reduce contaminants. Usually a filamentous fungus requiring aerobic condition is used and the inoculum is mixed into substrate for fermentation. SSF processes are those microbiological processes in which growth and product formation takes place on and inside the humidified solid substrate (Machado et al., 2004). In this case, processes occur in the absence or near absence of free water. There are four interacting phases present in such processes which include (i) a gas phase, (ii) a solid insoluble support, (iii) a liquid phase containing dissolved substrates and products, and (iv) a biotic phase formed by the microorganisms. Based on the type of solid substrate used, SSF systems have been divided into: (i) those employing natural material as solid substrate, (ii) those employing inert support impregnated with liquid medium (Nigam et al., 2009). Different agro-industrial wastes such as cassava bagasse, sugarcane bagasse, sugarbeat pulp/husk, orange bagasse, oil cakes, wheat bran, coir pith, coffee pulp/husk, okara etc are used as natural solid substrates, selection of which depends on some physical parameters such as particle size, moisture level, intra-particle spacing and nutrient composition within the substrate (Nigam et al., 2009). Madhumithah et al., carried out a study to produce protease from vegetable waste by SSF with *Aspergillus niger*. They used various vegetable wastes like potato, pumpkin, cauliflower, cabbage, and brinjal as a substrate for fermentation. They found maximum protein content i.e., 291.54 μg mL−1 from cauliflower among all the substrates. Similarly, Dhanasekaran et al., (2011) obtained single cell protein using pineapple residues as a raw material through two different strains of yeasts, *Saccharomyces cerevisiae*, and *Candida tropicalis*. The biomass contents increase with the increase in the concentration of pineapple waste. The highest biomass and protein content was observed on the 7th and 3rd day of fermentation with both the yeasts. SSF was also used for the extraction of lycopene using tomato waste as substrate by Jamal et al., (2016). Lycopene is an eminent carotenoid, producing the red colour of tomatoes and used as an antioxidant agent, colouring agent in the cosmetics, pharmaceutical as well as food industries. The effect of SSF by using fungus strain i.e., *Rhizopus oryzae* on the release of phenolic contents from rice bran was studied by Schmidt et al., (2014). A transformation in phenolic contents was observed, whereas ferulic acid giving the maximum enhancement during fermentation, initially from 33 mg/g in rice bran and finally 765 mg/g in the fermented bran.

Applications of SSF: Some of the industrial processes where SSF has successfully been applied include: i. Enzyme production – Some of the reported enzymes produced through SSF are protease, lipase, cellulose, pectinase, phytase, L-glutaminase, amylase, ligninase, xylanase, etc. ii. Organic acid production – Some of the organic acids produced via SSF includes lactic acid and gallic acid (Shojaosadati and Babaeipour 2002; Vandenberghe et al., 2000; Miura et al., 2004; John et al., 2006; Banerjee et al., 2005; Kumar et al., 1987). iii. Secondary metabolite production – There are various secondary metabolites which have been produced using SSF. Some of those products include gibberellic acids (Gelmi et al., 2000; Pastrana et al., 1995; Medeiros 2001), aroma production (Feron et al., 1996; Longo et al., 2006; Dominguez et al., 2001) , antibiotics (Phae et al., 1990; Sekar and balaraman 1998). iv.

Poly-gamma glutamate production (Oliviera 2004) v. Poly unsaturated fatty acid production (Vrije et al., 2001) vi. Biocontrol agent production (Larena et al., 2005; Pandey et al., 2004).

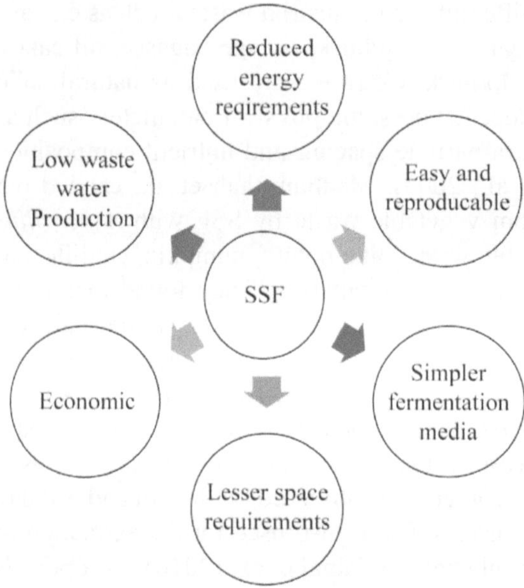

Fig. 2: Showing the advantages of fermentation (Stanbury et al., 1999)

Submerged Fermentation (SmF)

Submerged fermentation is the cultivation of microorganisms in liquid nutrient broth. Industrial enzymes/other products can be produced using this process. This involves growing carefully selected microorganisms (bacteria and fungi) in closed vessels containing a rich broth of nutrients (the fermentation medium) and a high concentration of oxygen. Production medium is required for growth of microorganism as well as production of primary and secondary metabolites. Submerged fermentations have been mainly used to produce flowable formulations. Contrary to SSF, submerged fermentation (SmF) are those microbiological processes in which growth and product formation take place using substrate present in liquid form. There are three main modes of SmF – (i) Batch mode, (ii) Fed-Batch mode, (iii) Continuous mode. Most industrial fermentation processes, except a few, operate as simple batch or fed-batch process.

- **Batch mode** – Batch culture is the simplest mode of operation. Here, initially all the nutrients required for the organism's growth and product formation are added in one vessel at the start of the fermentation process. This is followed by medium sterilization and inoculation of the vessel with desired organism for growth and product formation. This mode of operation operates as closed system and in between nothing is added into the vessel, except air. Finally, fermentation is terminated when either all the nutrient is exhausted or the desired concentration of product is achieved.

- **Fed-batch mode** – This is similar to batch process, except that these do not operate as closed systems. In this mode, one or more substrates, nutrients, and/or inducers are added in between during the process. This mode is normally used to extend the productive phase of the process.
- **Continuous mode** – In this mode of operation, the organisms are fed with fresh nutrients while removing the spent medium and cells at same rate so as to ensure that factors like culture volume, biomass, product and substrate concentrations along with pH, temperature and dissolved oxygen are constant during the process.

Pectinase, an enzyme production from fungi has been described by Favela-Tores et al., (2006) using SmF. Pectinases are a gathering of related proteins engaged with the breakdown of pectin from an assortment of plants. Pectinases have various commercial as well as industrial importance. Pectinase production also reported by Beg et al., (2000) ; Debing et al., (2006) and Biz et al., (2016) through fermentation processes using agro-wastes. Buyukkileci et al., (2015) produced a high amount of enzyme i.e., exo-polygalacturonase from orange peel by using SmF with *Aspergillus sojae*. Corn husks were used as a substrate for the production of cellulose from *Bacillus cereus* by SmF (Nema et al., 2015). SmF was performed for the amylase production using date wastes as a substrate with *Candida guilliermondii*. The process parameters such as incubation time, incubation temperature, initial pH, starch concentration, supplementary nitrogen source, nitrogen, and phosphorus concentrations that affect the production of α-amylase were also optimized (Acourene et al., 2013). Similarly, Budihal and Asgar (Budihal et al., 2015) performed the production of cellulase enzyme by Streptomyces DSK29 under both types of bioprocesses such as submerged and solid state fermentation using agro wastes as substrates.

Table 1: Showing comparison between Solid-state fermentation (SSF) and Submerged fermentation (SmF).

Parameter	SSF	SmF
Nature of substrate	Solid	Liquid
Volumetric Productivity	Relatively low	High
Energy requirements	Relatively high	Low

Aeration control	Relatively difficult to meet the requirements	Relatively easier to meet the requirements
Downstream processing	Relatively difficult	Relatively easier
Risk of desiccation of microbe used	High	Low
Inoculum used	Inoculum sprayed on surface medium	Inoculum usually in liquid form
Temperature used	Difficult to achieve	Relatively easier
pH control	Difficult to provide	Not overly difficult
Availability of nutrients to the microbes	Cannot be controlled to narrow limits if needed by the process	Can be controlled to narrow limits if needed by the process
Availability of oxygen to the biomass	Cannot be controlled at a particular level of saturation	Cannot be controlled at a particular level of saturation
Contamination problem	Relatively High	Low
Power consumption	Relatively Low	High
Foaming	No	Yes

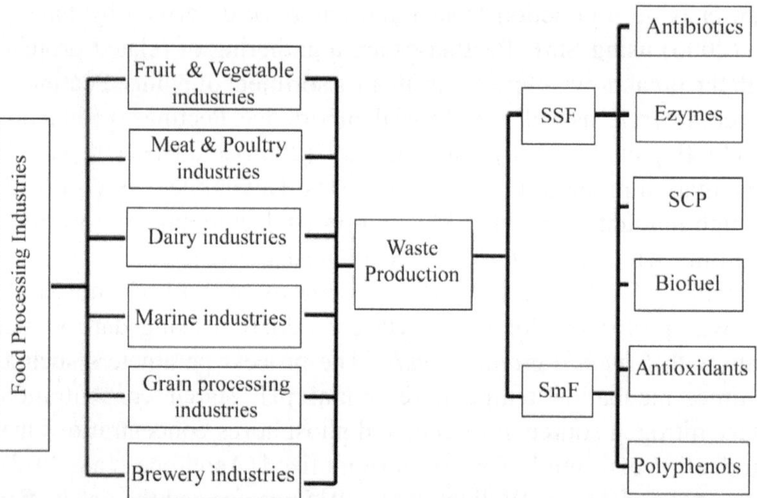

Fig. 3: Production of value added compounds from food industries waste using fermentation.

Prospects of fermentation technology

Depending on the type of fermentation different products such as proteins, enzymes, alcohol, acids, etc. are produced as shown in Fig 4.

Fermentation products			
	Microbial cells	**Yeast cells:** used in the baking industry **Single cell protein(SCP):** used as food material	
	Enzynes	**Bacterial:** amylase, protease, lipase. **Fungal:** amylase, protease, pectinase.	
	Metabolites	**Primary meabolites:** Proteins, vitamins, lipids, carbohydrates. **Secondary metabolites:** Enzyme, Antimicrobial compounds, growth promoters.	
	rDNA products	**Hormones:** Growth hormone, insulin. **Therapeutic proteins:** clotting factors-VIII&IX, interferons.	
	Transformation products	**Biofuels production:** Bioethanol, biohydrogen, bioplastic **Bioremediation:** Wasterwater, aromatics, toxic dyes.	Biofuel

Some of the industrially valuable fermentation products are discussed below:

Biofuel Production

In recent years increase in need for biofuels in replace of fossil fuels are in high demands. Fermentation is one of the most important step followed to generate fossil fuels from lignocellulosic waste or other biodegradable sources such as microalgae etc. Biofuels such as biodiesels, bioethanol, butanol, biohydrogen etc. are one of the most important value added products which are produced from biodegradable and waste materials leading to zero waste concept.

- **Biodiesel:** Vegetable oil or animal fats or any natural form of oil after trans esterification produces biodiesel. Trans esterification is a process of exchanging organic group of an ester with the organic group of an alcohol under catalysis of acid or base. Fatty acids, triglycerides or oils are found from many plant, animal or microbial resources Demirbas 2011. Many process are involved to extract oils from lignocellulosic biomass which on further trans-esterified to form biodiesel. Though till now plant based oils are most abundantly used for biodiesel production but it has some limitation. Plant based oils are also used as edible sources thus, raising conflict on providing priority on production of biofuels or used as foods. Other than this some non-edible oil rich crops are present but less abundance of lands for non-edible crops raise the problem (Arazo et al., 2017; Milano et al., 2016). Microalgae in such case is the solution. Many researches are done on microalgae for the production of biodiesel. Depending on species 70% of the cell consist of triglyceride in case of microalgae (Milano et al., 2016). Microalgae can be easily cultured and maintained even in hash condition and in large quantity. Fermentation of microalgae produces bio-oil which on further transesterification produces biodiesel. As microalgae grows very fast thus cultivating microalgae in fermenter solve the problem of space as well as conflict on food crop. Heterotrophic fermentation of Chlorella protothecoides on glycerol as carbon source using semi continuous fermentation process produced more bio-oil than fedbatch fermentation thus, yielding high bio-oil for biodiesel production (Garcia et al., 2013). Being scarcity of fossil fuel and its high need, production of biodiesel from these renewable sources is a relief for the world, leading to rapidly growing industries in this sector.

Fig. 5: Schematic representation of biodiesel production through algae. (Fatih 2010)

- **Enzymes**

Enzyme cultivation is the most important technique for the manufacturing of different enzymes. When fermentation on appropriate substrates is done, both fungus and bacterial microbes are required for the precious collection of enzyme. Enzyme production can be together performed by submerged and SSF. SSF is commercially used in Japan to produce industrial enzymes. Initial investigations at Biocon India Limited (Bangalore, India), more than 15 years ago, comparing minor metabolite profiles of similar enzymes produced by

SSF and by SmF led to the conclusion that commercially available solid-state produced enzymes were richer in side activities compared to commercially available SmF enzymes, when normalized to the same main activity. In the food-processing industry, most substrates for enzymes are naturally occurring raw materials, like cereals, fruits or vegetables, and since these materials are always composed of a variety of complex ingredients like starch, xylans, cellulosics, proteins, and fats, it was quite understandable that solid-state-produced enzymes do a better job of converting these raw materials to give higher yields. Biocon India, therefore, began to focus on the production of enzymes by SSF.

α-amylases (endo-1,4-β-D-glucan glucohydrolases EC 3.2.1.1) are extracellular enzymes secreted as a primary metabolite that randomly cleave the 1,4-α-D glucosidic linkages between adjacent glucose units in the linear amylose chain. The extensive application of amylases in the food industry, such as baking, brewing, preparation of digestive aids, production of chocolate cakes, moist cakes, fruit juices, starch syrups, has paved a way for their large-scale commercial production. Generally, the production has been carried out using SmF however, SSF systems appear promising due to the natural potential and advantages they offer (Pandey et al., 1999). The production of α-amylase by SSF is limited to the genus Bacillus and *B. subtilis* , *B. polymyxia* , *B. mesentericus, B. vulgarus, B. megaterium and B. licheniformis* have been used for α-amylase production in SSF (Babu et al., 1995). The production of bacterial α-amylase using the SSF technique requires only a fermentation time of 24–48 h (Ramesh et al., 1987). Thus, with reduced batch time, the use of SSF technique for bacterial α-amylase production leads to a considerable reduction in capital and recurring expenditure. Mulimani and Ramalingam (2000) tested several substrates (rice straw, rice bran, rice straw + rice bran, red gram husk, jowar straw, jowar spathe, wheat bran) for the production of α-amylase by the fungus *Gibberella fujikuroi* in SSF. Among the tested substrates, wheat bran led to the maximum α-amylase production. Francis et al., 2003 studied the production of α-amylase by *A. oryzae* in SSF using spent brewing grains as the sole carbon source. Under optimised conditions (30°C, 70% initial moisture and an inoculum rate 1 × 107 spores/g ds) about 20% increase in enzyme yield was observed. Ramachandran et al., (2004) used coconut oil cake as a substrate for the production of α-amylase by A. oryzae under SSF conditions. They obtained 1372 U/g ds α-amylase in 24 h.

Sabu et al. (2008) showed that palm kernel cake and tamarind seed powder, abundant wastes in many tropical and subtropical countries, were very suitable as substrates for the production of tannase under SSF conditions by *A. niger.* Thus, the former led to 13.03 IU/g ds and the latter to 6.44 IU/g ds. Recently,

Kumar et al., (2007) studied the production of tannase by *Aspergillus ruber* under SSF using different tannin-rich substrates like ber leaves (*Zyzyphus mauritiana*), jamun leaves (*Syzygium cumini*), amla leaves (*Phyllanthus emblica*) and jawar leaves (*Sorghum vulgaris*). Jamun leaves were found to be the best substrate for tannase production, leading under optimised conditions to 69 U/g ds.

Antibiotics

The most important extract from microorganism using fermentation technology is antibiotics. It is a bioactive compound. Penicillin from *Penicillium notatum* is the first antibiotic produced from fermentation. It was completed in 1940s using SSF and SmF but today *P. chrysogenum* isolates are higher yielding producers (Balakrishnan and Pandey 1996). Aminocillins, Carbapencins, Monobactams, Cephalosporins and Penicillins together they are known as β-lactam antibiotics (Balakrishnan and Pandey 1996). Some other antibiotics like Tetracyclin, Streptomycin, Cyclosporin, Cephalosporin and Surfactin are manufactured from this process. *Streptomyces clavuligerus*, and *Streptomyces cattleya* produces Cephamycin C from sunflower cake and cotton-de-oiled cake in which wheat raw is supplemented in SSF system as substrates for manufacturing Cephamycin C. In SSF, penicillin was produced by actinomycetes and fungi in mixed cultures. In current time, the growth of proper substrates has led to the widespread use of SSF more than SmF. On the other hand, some results show that several microbial stains are extra suitable to SSF and others are more suitable for SmF. Thus, this technology is determined on the basis of microorganism that is being used for production. Recently, it has been studied that several antibiotics produced through SSF are more constant and high in quantity than SmF. However, the characteristics of the substrate material and their quality make SSF implementation limited. Due to this property, it is compulsory to check the production ability of different substrates earlier than optimization of the fermentation process.

Ellaiah et al., (2004) tested several support-substrates (wheat bran, wheat rawa, rice bran, rice rawa, rice husk, rice straw, maize bran, ragi bran, green gram bran, black gram bran, red gram bran, corn flour, jowar flour, sago and sugar cane bagasse) for neomycin production by a mutant strain of *Streptomyces marinensis*, under SSF. Among the different substrates tested, wheat rawa was found to be the most suitable for neomycin production. A maximum neomycin yield of 10 755 μg/g substrate was achieved under optimised conditions.

This value was 140% higher than that obtained by the control culture. In addition, earlier studies with SmF using a synthetic medium showed 5813 μg/mL neomycin yield. Therefore, the accumulation of neomycin by SSF was 1.85 times higher than the SmF. Asagbra et al., (2005) assessed the ability of Streptomyces sp. OXCI, S. rimosus NRRL B2659, *S. rimosus* NRRL B2234, *S. alboflavus* NRRL B1273, *S. aureofaciens* NRRL B2183 and *S. vendagensis* ATCC 25507 to produce tetracycline under SSF conditions using peanut shells, corn cob, corn pomace and cassava peels as substrates. They found that peanut shells were the most effective substrate. Mizumoto et al., (2006) reported the production of the lipopeptide antibiotic Iturin A by *B. subtilis* using soya bean curd residue, okara, a by-product of tofu manufacture in SSF. After 4 days of incubation, iturin A production reached 3300 mg/kg wet solid material (14 g/kg ds material), which was approximately tenfold higher than that in SmF.

Organic Acids

Lactic Acid

One of the most widely used organic acid by pharmaceutical, food, and textile industries is Lactic acid (2-hydroxypropionic acid). In food industries, it is employed as an acidulant, preservative, and flavoring agent. Lactic acid found its use as a raw material in the manufacturing of propylene glycol, lactate ester, 2,3-pentanedione, acrylic acid, propanoic acid, dilactide, and acetaldehyde (Wee et al., 2006; Varadarajan and Miller 1999) . Nowadays, the demand of lactic acid has increased manifolds all over the world because it can be used in the synthesis of PLA (polylactic acid). PLA is a biopolymer through which many biodegradable polymers or bioplastics can be manufactured. The two optical isomers of lactic acid i.e., L (+)- lactic acid and D (–)-lactic acid is very well known (Datta et al., 1995). The optical nature of lactic acid is very important factor for synthesizing a premium PLA . Lactic acid can either be produced chemically or through the process of fermentation (Lunt 1998). However, fermentation is preferred more for the production as this process synthesizes the optically pure lactic acid. On the other hand, a racemic mixture of D and L lactic acid is produced when manufactured chemically (Ghaffar et al., 2014). The petrochemical resources are used as a substrate for producing lactic acid through chemical process while in biological method, sugar molecules occurring in biodegradable material such as lignocellulosic biomass gets fermented by microorganisms to form lactic acid (Akerberg and Zacchi 2000). All the substrates in fermentation process are biodegradable and renewable, therefore, this method is considered as eco-friendly. For synthesizing lactic acid through the process of fermentation many waste materials like agro-waste (Chaisu et al., 2014), waste obtained from food industries (Pleissner et al.,

2017), wastewater from Cassava starch (Tosungneon et al., 2014), and many other lignocellulosic waste products can be employed. There are two types of lactic acid bacteria: homofermentative and heterofermentative .

The sugar is completely converted into lactic acid by homofermentative bacteria whereas glucose gets catabolizes into lactic acid (main product) and carbon dioxide as well as ethanol (by product) by heterofermentative bacteria. The lactic acid gets synthesized by homofermentative bacteria through the process known as glycolysis pathway or Embden-Meyerhof pathway (Thomas et al., 1979). Some of the microbes known for the production of lactic acid are *Rhizopus oryzae* NRRL 395, *Rhizopus oryzae* ATCC 52311, *Lactobacillus rhamnosus* ATCC 10863, *Enterococcus faecalis* RKY1, *Lactobacillus bulgaricus* NRRL B-548, *Lactobacillus helveticus* ATCC 15009, *Lactobacillus plantarum* ATCC 21028, *Lactobacillus casei* NRRL B-441, *Lactobacillus amylophilus* GV6, *Lactobacillus pentosus* ATCC 8041, *Lactococcus lactis* sp. IFO 12007 *Lactobacillus delbrueckii* NCIMB 8130. Chaisu et al. (2014) used *Lactobacillus casei* M-15 for optimizing the production of lactic acid by using agricultural byproduct such as molasses obtained from sugarcane . The maximum yield of lactic acid was optimized by growing the culture in 3.82% molasses along with 8.02% inoculum at 37 °C for 24 hours using RSM (Response Surface Methodology). The production of lactic acid significantly increased through this process and also follows sustainable development. Also, this process helps in managing the waste by utilizing the agricultural waste products and is free from pollution. Another study was conducted by Tosungnoen et al. (2014) in which they treated the amylolytic *Lactobacillus plantarum* MSUL 702 culture with the wastewater of synthetic cassava starch (SCW). They followed the process of repeated-batch and simultaneous saccharification and fermentation for producing lactic acid. The first batch of culture synthesized 28.71 g/L of lactic acid. Also, the second batch of culture reduced the 85% of TKN (total kjeldahl nitrogen) and 98% of COD (chemical oxygen demand). This culture produces maximum lactic acid and also retains the efficiency of treatment for four consecutive batches, thereby reusing the waste product for forming value-added products . For producing lactic acid, Pleissner et al. (2017) used food waste obtained from restaurants . Food waste have a lot of nitrogen and sugar resources that is used for producing L (+)-lactic acid through simultaneous fermentation and saccharification by employing *Streptococcus* or *Lactobacillus* sp. Under sterile condition and technical scale, 702 g/l of purified L (+) lactic acid was produced in the laboratory. Therefore, the household waste, restaurant waste, or industrial waste can be employed for synthesizing important industrial materials.

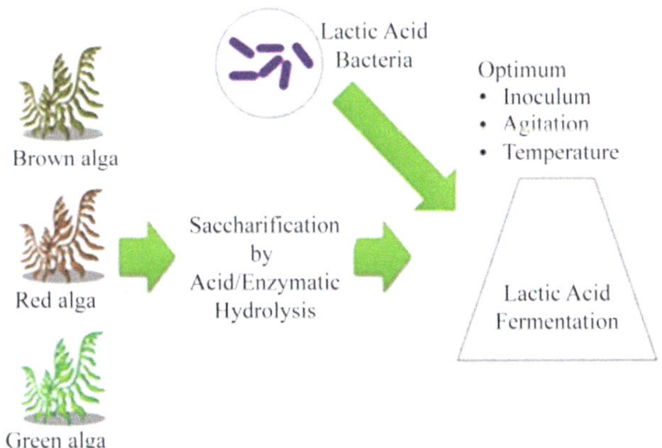

Fig. 6: Production of Lactic Acid from Seaweed Hydrolysates via Lactic Acid Bacteria Fermentation

Polyhydroxyalkanoates

Polyhydroxyalkanoates or PHA are naturally-occurring hetero and homo polysaccharides such as poly 3-hydroxybutyric-co-3-hydroxyvaleric acid (PHB-V) and poly 3-hydroxybutyric acid (PHB). Many microbes such as *Alcaligenes latus, methylotrophs, Azotobacter vinelandii, Ralstonia eutropha, Chromobacteruium violaceum, and pseudomonas* produced these in very limited nutrient content (Madison and Huisman 1999). In the limited amount of any one of the nutrients such as phosphorus, sulfur, nitrogen, oxygen, or magnesium and surplus carbon sources, these microbes produce PHA within their cells as an energy source and stored carbon. The microorganisms easily break the PHA into water and carbon dioxide. PHA is a lactic acid polymer which acts as a green polymer for synthesizing bioplastics. The microorganisms produce two types of PHA, first one is short-chain-length PHA (scl-PHA) which have three to five carbon atoms. The second form of PHA is medium-chain-length PHA (mcl-PHA) which is made up of six to fourteen carbon atoms. A category of scl-PHA is brittle and stiff and called as PHB while PHB-V is another type of PHA which is formed by combining 3-hydroxyvalerate (HV) and PHB resulting in a flexible, tougher and thermally stable PHA (Lopez et al., 2015). The use mcl-PHA generally resides in the manufacturing of elastomers while sci-PHB is mostly employed as biodegradable packaging material. PHA has now become the most important product of various industries as bio polymers and bioplastics have high demand in the market. But because of its high production cost, the PHA is not widely industrialized and commercialized (Kourmentza et al., 2017). Many PHA-synthesizing bacteria can survive under stressful conditions. Many bacterial strains belonging to the genus *Bacillus, Sphingobacterium, Rhodococcus* and *Pseudomonas*, produces PHA and helps in degrading the pollutants present in the environment . Most of the environmental pollutants emit VOC (volatile organic compounds) that are harmful for the human beings and can cause carcinogenic, mutagenic, cytotoxic, and allergic reactions. The pollution has nowadays increased very much because of the over use of petrochemicals. In this regard, the PHA is recognized as a valuable product because the bacteria synthesizing PHA can disintegrate these hydrocarbons in the process and therefore, acts as a boon for the environment (Khandare et al., 2015). PHA can be synthesized from the mixture of many organic compounds such as ethylbenzene, benzene, styrene, and xylene by some *Pseudomonas* sp. like *P. putida* F1, CA-3, and mt-2 (Nikodinovic et al., 2008). As vegetable oils are enriched with carbon sources therefore, used as a substrate for the synthesis of PHA. The discussion on the production of PHA was reported by Hwan et al. (2008). They used the bacteria *Pseudomonas* sp. strain DR2 and vegetable oil as a substrate for successfully synthesizing the PHA (Song 2008). Rhu et al. (2003) reported the sewage waste management and PHA synthesis. As PHA-producing bacteria can tolerate stress, therefore, they can be employed for treating the hydrocarbon-rich waste generated by sewage water . Another study conducted by Colombo et al. (2017) in Europe reported the synthesis of PHA by using a variety of PHA-synthesizing microorganisms and municipal waste as a substrate (Columbo 2017). The bacteria produce PHA with the help of an enzyme called PHA synthase. The bacteria *P. palleronii* and *P. pseudoflava* has the ability to disintegrate the synthetic waste while producing PHA. These bacteria were used for isolating the enzyme PHA synthase by Reddy et al. (2017) which can be employed for waste management and PHA production in a pocket-friendly way. Therefore, synthesizing PHA by using waste resources decreases the production cost which is helpful in bioremediation. This is an effective way to convert the waste product into value-added industrial material.

Exopolysaccharides (EPS)

Exopolysaccharides (EPS) are synthesized by different bacteria from sugar molecules. They are of two types, homopolysaccharides and heteropolysaccharides. Homopolysaccharides such as dextran, alternan, pullulan, levan, β-D-glucansare synthesized by bacteria *Leu. mesenteroides* subsp. mesenteroides, *Leu. mesenteroides, Aureobasidium pullulons, levan, Z. mobilis,* Streptococcus sp. respectively. Hetropolysaccharides such as alginate, gellan, xanthan are synthesized by bacteria *Pseudomonas aeruginosa, Sphingomonas paucimobilis, Xanthomonas campe* respectively. Other than these EPS are also synthesized by lactic acid bacteria of different chemical structures. EPS synthesized by lactic acid bacteria provides consistency, texture, and rheology of the fermented milk products. Synthesis of microbial EPS is very complex process and involves lots of genes. Generally, synthesis of EPS involved four major steps which are: import of sugar into the cytoplasm, synthesis of sugar-1-phosphates, activation and coupling of sugar molecules, export of EPS from cytoplasm to media. EPS has numerous application in food industries, medical, and pharmaceuticals (Laws et al., 2001). EPS producing bacteria are also known to grow on waste water and acts as waste treatment. Some of the incidences are discussed here. Taskin et al. (2011) has reported about synthesis of EPS by *Morchella esculenta* on extract of waste loquat *(Eriobotrya japonica* l.) kernels. EPS production on an average of 5 g/L was recovered (Taskin et al., 2011). *Shochu* is popular beverage in Japan which is produced from rice, sweet potatoes, barley, buckwheat, or sugar beets, shochukasu which is the distillery waste of shochu production is prohibited to dump in river or fields. Thus, it is dumped in ocean which is costlier as well as causes marine pollution. As this waste is rich in sugar contained, thus, Yuliani et al. (2011) used this wastewater to culture *Lactobacillus sakei* CY1 and produced EPS of 25.5% glucose and 13.2% galactose along with other sugars. Morillo et al. (2007) isolated a bacterial strain *Paenibacillus jamilae* from olive mill wastewater which produces EPS. This bacterium produces 5.1 g/L of EPS from this wastewater using batch culture fermentation method. Sellami et al. (2015) synthesized EPS by cultivating *Rhizobium leguminosarum* on waste water of oil company and wastewater of fish processing company. The highest EPS production of 42.4 g/L, after 96 h of culture was found when both oil and fish processing company wastewater was mixed. Thus, the discussion emphasizes that EPS which is one of the valuable industrial product can be synthesized efficiently from waste biomass and also help in bioremediation of wastes (Sellami 2015)

Flavoring and Fragrance Substances

There are varieties of flavoring and fragrance substances produced by chemical process and are important for industrial use especially in food and cosmetic industries. Flavoring substances are generally volatile and non-volatile chemicals which are extracted from plant sources or synthesized chemically (Sanroman 2006). Volatile chemical only contribute to the aroma but non-volatile chemicals contribute both to the aroma and taste . Flavors are generally extracted from plant sources and then their chemical structure is studied and chemical flavors are synthesized. The aroma of the food is responsible for some compounds such as alcohols, aldehydes, esters, dicarbonyls, short to medium-chain free fatty acids, methyl ketones, lactones, phenolic compounds, and sulphur compounds (Urbach 1997). Though flavors can be extracted from plant sources but the downstream process for such extraction are very costly and also very small quantities of compound can be extracted from a large biomass. Thus, flavors and fragrances are chemically mimicked. But this chemical process of flavoring substance causes environmentally unhealthy production and lacks substrate selectivity which causes undesirable racemic mixtures, thus increasing the cost of processing. As natural fragrances and flavoring substances are in high demand thus, microbial technology are used for the production of safe and natural fragrance and flavoring substances. Microorganism can synthesize these compounds as secondary metabolites on nutrients such as sugars and amino acids. This can be done by two process, one by in situ production of aroma and flavor as integral part of food and beverage production process and another is by synthesizing aroma or flavors from specifically designed microbial system. Some of the example of flavors and fragrance synthesis from waste biomass as substrate are discussed. Phenylethyl alcohol which is aroma of rose can be synthesized by Kluyveromyces marxianus from cheese waste (Conde- Baez 2017). Terpenoid flavor and fragrance compounds are one of the most industrial valuable compounds which could be synthesized by microorganism fermentation of house- hold vegetable wastes (Schempp 2017). *Rhizopus oryzae* and *Candida tropicalis* has the ability to synthesize D-limonene a flavoring substance from olive oil mill waste (Fatih 2010).

Table 2: Production of different industrially important metabolites by using varied substrate

Substrate	Micro-organisms	Product
Spent malt grains, apple pomace, grape, citrus peels	*Xanthomonas campestris*	Xanthan (Stredansky et al., 1999)
Apple, orange and potato peelings	*Trametes hirsuta*	Laccase (Rosales et al., 2002)
Sugarcane bagasse, wheat bran, molasses	*A. niger*	Citric acid (Vanderberghe et al., 2000)
Wheat bran	*Aspergillus parasiticus*	Protease (Tunga et al., 2003)
Grape pomace	*Aspergillus awamori*	Xylanase, pectinase (Botella et al., 2007)
Cassava bagasse, apple pomace	Rhizopus (four different strains	Volatile carbons as flavours, (Christen et al., 2000) acetaldehyde, ethanol, propanol, esters
Sweet potato residue	*Streptomyces viridifaciens*	Tetracycline (Yang and Ling 1989)
Wheat bran, rice husk	*Bacillus subtilis*	α-Amylase (Baysal et al., 2003)
Solid wheat bran and 10% purified chitin flakes	*Penicillium aculeatum, Penicillium chrysogenum,*	Chitinase (Binod et al., 2005)

Conclusion

Fermentation technology is use of microbes for the production of various industrially valuable products (Fig 7). Optimization of fermentation technology provide green route for production of various chemicals, enzymes, proteins, biofuels, polymers, etc. with low cost of production and no pollution. Nowadays, many synthetic products are used which are derived from petrochemical industries and are harmful for human health and environment. Through fermentation technology, chemicals which mimics the chemical produced by petrochemical process can be developed. Moreover, with fermentation technology waste products of the industries can be converted to valuable products leading to zero waste technology. Biofuels can be easily produced from different industrial waste and similarly different chemicals, polymers, enzymes, flavoring substance are produced by fermentation process using waste materials as substrate. This not only reduces the cost of production but also help in bioremediation. Some of the polymers such as PHA and PLA are used for the production of bioplastics and replace the petrochemical-based plastics. Other than this hormones and enzymes production were costlier earlier, but with the help of recombinant DNA technology, genetically modified microorganism are capable of producing those enzymes and hormones by fermentation technology. Exopolysaccharides are secreted by bacteria which are highly valuable for pharmaceutical industries. Whereas ligninolytic enzymes like laccase and cellulase are valuable for paper, textile, and biofuel industries. This biofuel itself is a big relief to the environment as it is supplement to fossil fuels. Thus, industrially value added product formation and converting waste as valuable raw material has been discussed in this chapter. More research is furnishing with this enormous microbial diversity for green technology development and producing every possible product naturally and safe for the environment.

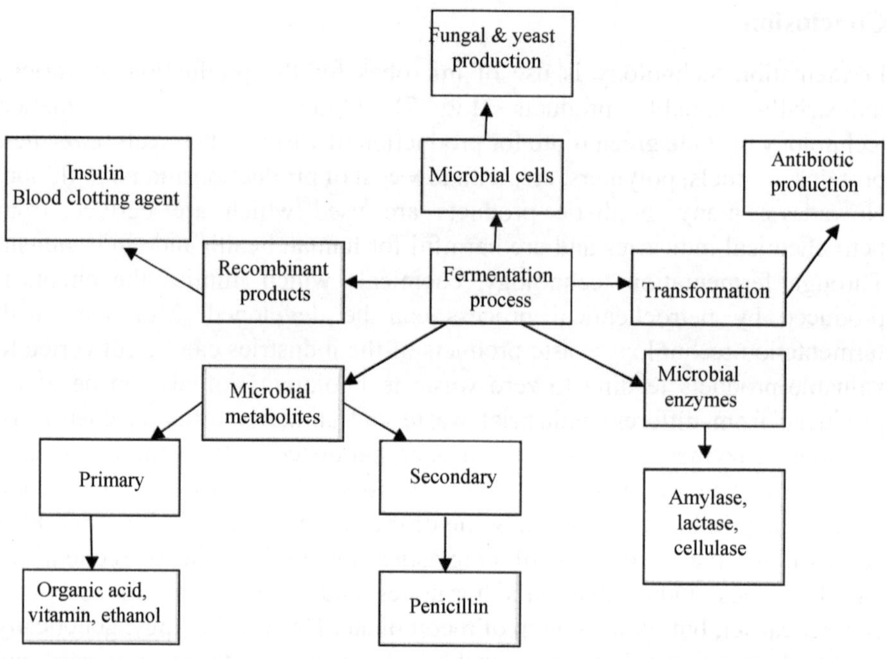

Fig. 7: Important products derived via the fermentation process (Stanbury et al., 2016)

References

A. Pandey, F. Francis, A. Sabu, C.R. (2004) Soccol, Concise Encyclopedia of Bioresource Technology, p. 702.

A.B.P. Medeiros, A. Pandey, P. Christen, P.S.G. Fontoura, R.J.S. Freitas, C.R. Soccol, (2001) World Journal of Microbiology and Biotechnology, Vol. 17, p. 767.

A.E. Asagbra, A.I. Sanni, O.B. Oyewole, (2005). World Journal of Microbiology and Biotechnology, Vol. 21, p. 107.

Acourene, S.; Amourache, L.; Benchabane, A.; Djaafri, K. (2013). Utilisation of date wastes as substrate for the production of α-amylase. Int. Food Res. J. 20, 1367–1372.

Åkerberg, C. and G. Zacchi, (2014) An economic evaluation of the fermentative production of lactic Chaisu, K., et al., Optimization Lactic Acid Production from Molasses Renewable Raw Material through Response Surface Methodology with Lactobacillus Casei M-15. APCBEE Procedia, 8(Supplement C): p. 194–198.

Aradarajan, S. and D.J. Miller, (1999) Catalytic Upgrading of Fermentation-Derived Organic Acids. Biotechnol Prog, 15(5): p. 845–854.

Arazo, R.O., de Luna, M.D.G., Capareda, S.C., (2017) Assessing biodiesel production from sewage sludge-derived bio-oil. Biocatal. Agric. Biotechnol., 10(Supplement C), 189–196.

Asagbra, A. E., Sanni, A. I., Oyewole, O. B., (2005) Solid-state fermentation production of tetracycline by Streptomyces strains using some agricultural wastes as substrate. World J. Microb. Biotechnol. 21, 107–114.

Babu, K. R., Satyanarayana,T., (1995) Amylase production by thermophilic Bacillus coagulans in solid state fermentation. Process Biochem. 30, 305–309.

Balakrishnan, K., Pandey, A., (1996) Production of biologically active secondary metabolites in solid state fermentation. J. Sci. Ind. Res., 55, 365.

Baysal, Z., Uyar, F., Aytekin, C., (2003) Solid state fermentation for production of α-amylase by a thermotolerant Bacillus subtilis from hot-spring water. Process Biochem, 38, 1665–1668.

Beg, Q.K.; Bhushan, B.; Kapoor, M.; Hoondal, G.S. (2000) Enhanced production of a thermostable xylanase from Streptomyces sp. QG-11-3 and its application in biobleaching of eucalyptus kraft pulp. Enzyme. Microb. Technol. 27, 459–466. [CrossRef]

Binod, P., Sandhya, C., Pusztahelyi,T., Szakacs, G. et al., (2005) Production and purification of extracellular chitinases from Penicillium spp. under solid state fermentation. Enzyme Microb. Tech. 36, 880–887.

Biz, A.; Finkler, A.T.J.; Pitol, L.O.; Medina, B.S.; Krieger, N.; Mitchell, D.A. (2016) Production of pectinases by solid-state fermentation of a mixture of citrus waste and sugarcane bagasse in a pilot-scale packed-bed bioreactor. Biochem. Eng. J. 111, 54–62. [CrossRef]

Botella, C., Diaz, A., de Ory, I., Webb, C., Blandino, A., (2007) Xylanase and pectinase production by Aspergillus awamori on grape pomace in solid state fermentation. Process Biochem. 42, 98–101.

Budihal, S.R.; Agsar, D. (2015) Exploration of Agrowastes for the Production of Cellulase by Streptomyces DSK29 under Submerged and Solid State Systems. Int. J. Curr. Microbiol. Appl. Sci. 4, 681–689.

Buyukkileci, A.O.; Lahore, M.F.; Tari, C. (2015) Utilization of orange peel, a food industrial waste, in the production of exo-polygalacturonase by pellet forming Aspergillus sojae. Bioprocess Biosyst. Eng. 38, 749–760. [CrossRef] [PubMed]

C. Gelmi, R. Pér ez-Correa, M. González, E. Agosin, (2000) Process Biochemistry, Vol. 35, p. 1227.

C. Sekar, K. Balaraman, (1998) Bioprocess Engineering, Vol. 18, p. 293.

C. Sekar, V.W. Rajasekar, K. Balaraman, (1997) Bioprocess Engineering, Vol. 17, p. 257.

C.G. Phae, M. Shoda, H. Kubota, (1990) Journal of Fermentation and Bioengineering, Vol. 69, p. 1.

Cerón-García, M.C., et al., (2013) A process for biodiesel production involving the heterotrophic fermentation of Chlorella protothecoides with glycerol as the carbon source. Appl. Energy, 103(Supplement C), 341–349.

Christen, P., Bramorski, A., Revah, S., Soccol, C. R., (2000) Characterization of volatile compounds produced by Rhizopus strains grown on agro-industrial solid wastes. Bioresour. Technol. 71, 211–215.

Colombo, B., et al., (2017) Enhanced polyhydroxyalkanoate (PHA) production from the organic fraction of municipal solid waste by using mixed microbial culture. Biotechnology for Biofuels, 10(1): p. 201.

Conde-Báez, L., et al., (2017) Evaluation of Waste of the Cheese Industry for the Production of Aroma of Roses (Phenylethyl Alcohol). Waste and Biomass Valorization, 8(4): p. 1343–1350.

Datta, R., et al., (1995) Technological and economic potential of poly(lactic acid) and lactic acid derivatives. FEMS Microbiology Reviews, 16(2): p. 221–231.

Debing, J.; Peijun, L.; Stagnitti, F.; Xianzhe, X.; Li, L. (2006) Pectinase production by solid fermentation from Aspergillus niger by a new prescription experiment. Ecotoxicol. Environ. Saf. 2006, 64, 244–250. [CrossRef] [PubMed]

Demirbas, A., (2011) Competitive liquid biofuels from biomass. Appl. Energy, 88(1), 17–28.

Dhanasekaran, D.; Lawanya, S.; Saha, S.; Thajuddin, N.; Panneerselvam, A. (2011) Production of single cell protein from pineapple waste using yeast. Innov. Rom. Food Biotechnol. 8, 26–32

E. Cannel, M. Moo-Young, (1980) Process Biochemistry, Vol. 4, p. 2.

Ellaiah, P., Shrinivasulu, B., Adinarayana, K., (2004) Optimization studies on neomycin production by a mutant strain of Streptomyces marinensis in solid-state fermentation. Process Biochem. 39, 529–534

F.C. Oliveira, D.M.G. Freire, L.R. Castilho, (2004) Biotechnology Letters, Vol. 26, p. 1851.

Fatih DM (2010) Microalgae as a feedstock for biodiesel. Energ Educ Sci Technol A Energ Sci Res 25:31–43

Favela-Torres, E.; Volke-Sepúlveda, T.; Viniegra-González, G. Hydrolytic (2006) Depolymerising Pectinases. Food Technol. Biotechnol. 44, 221–227.

Francis, F., Sabu, A., Nampoothiri, K. M., Ramachandran, S. et al., Use of response surface methodology for optimizing process parameters for the production of α-amylase by *Aspergillus oryzae*. Biochem. Eng. J. 2003, 15, 107–115

Ghaffar, T., et al., (2014) Recent trends in lactic acid biotechnology: A brief review on production to purification. Journal of Radiation Research and Applied Sciences, 7(2): p. 222–229.

G. Feron, P. Bonnarame, A. Durand, (1996) Trends in Food Science and Technology, Vol. 7, p. 285.

Humer, E., and K. Schedle. (2016) "Fermentation of food and feed: a technology for efficient utilization of macro and trace elements in monogastrics." J Trace Elem Med Biol 37:69–77.

I. Larena, R. Torres, A. Cal, M. Li ñán, P. Melgarejo, P. Domenichini, A. Bellini, J.F. Mandrin, J. Lichou, X.O. Eribe, J. Usall, (2005) Biological Control, Vol. 32, p. 305.

Jamal, P.; Akbar, I.; Yumi, Z.; Irwandi, J. (2016) Process development for maximum lycopene production from selected fruit waste and its antioxidant and antiradical activity. J. Food Process. Technol. 7, 576–581. [CrossRef]

Khandare, R.V. and S.P. Govindwar, (2015) Phytoremediation of textile dyes and effluents: Current scenario and future prospects. Biotechnology Advances, 33(8): p. 1697–1714.

Kourmentza, C., et al., (2017) Recent Advances and Challenges towards Sustainable Polyhydroxyalkanoate (PHA) Production. Bioengineering, 4(2): p. 55.

Kumar, R., Sharma, J., Singh, R., (2007) Production of tannase from Aspergillus ruber under solid-state fermentation using jamun (Syzygium cumini) leaves. Microbiol. Res. 162, 384–390

L.M. Pastrana, M.P. Gonzalez, J. Pintado, M.A. Murado, (1995) Enzyme and Microbial Technology, Vol. 17, p. 784.

L.P.S. Vandenberghe, C.R. Soccol, A. Pandey, J.M. (2000) Le -bault, Bioresource Technology, Vol. 74, p. 175.

Laws, A., Y. Gu, and V. Marshall, (2001) Biosynthesis, characterisation, and design of bacterial exopolysaccharides from lactic acid bacteria. Biotechnol. Adv., 19(8): p. 597–625.

Lopez, N.I., et al., (2015) Polyhydroxyalkanoates: Much More than Biodegradable Plastics. Adv Appl Microbiol, 93: p. 73–106.

Lunt, J., (1998) Large-scale production, properties and commercial applications of polylactic acid polymers. Polymer Degradation and Stability, 59(1): p. 145– 152.

M. Dominguez, A. Mejia, S. Revah, J. Barrios-Gonzalez, (2001) World Journal of Microbiology and Biotechnology, Vol. 17, p. 751.

M.A. Longo, M.A. Sanromán, (2006) Food Technology and Biotechnology, Vol. 44, p. 335.

Machado, C.M., Oishi, B.O, Pandey, A., Socco, C.R., (2004) Kinetics of Gibberellafujikorigrowth andGibberellic acid production by solid state fermentation in a packed-bed column bioreactor. Biotechnol. Prog., 20, 1449.

Madhumithah, C.G.; Krithiga, R.; Sundaram, S.; Sasikumar, C.S.; Guhathakurta, S.; Cherian, K.M. (2011) Utilization of vegetable wastes for production of protease by solid state fermentation using Aspergillus niger. World J. Agric. Sci., 7, 550–555.

Madison, L.L. and G.W. Huisman, (1999) Metabolic engineering of poly(3-hydroxyalkanoates): from DNA to plastic. Microbiol Mol. Biol. Rev., 63(1): p. 21–53.

Martins, S., Mussatto, S.I., Martinez-Avila, G., Montanez-Saenz, J., Aguilar, C.N., Teixeira, J.A., (2011) Bioactive phenolic compounds: production and extraction by solid-state fermentation. a review. Biotechnol. Adv., 29, 373.

Milano, J., et al., (2016) Microalgae biofuels as an alternative to fossil fuel for power generation. Renewable Sustainable Energy Rev., 58(Supplement C), 180–197.

Mizumoto, S., Hirai, M., Shoda, M., (2006) Production of lipopeptide antibiotic iturin A using soybean curd residue cultivated with Bacillus subtilis in solid-state fermentation. Appl. Microbiol. Biotechnol. 72, 869–875.

Morillo, J.A., et al., (2007) Production and characterization of the exopolysaccharide produced by Paenibacillus jamilae grown on olive mill-waste waters. World Journal of Microbiology and Biotechnology, 23(12): p. 1705.

Motarjemi, Y., (2002) Impact of small scale fermentation technology on food safety in developing countries. Int. J. Food Microbiol., 75(3), 213–29.

Mulimani, V. H., Ramalingam, P. G. N., (2000) Amylase production by solid state fermentation: A new practical approach to biotechnology courses. Biochem. Educ. 28, 161–163.

Nema, N.; Alamir, L.; Mohammad, M. (2015) Production of cellulase from Bacillus cereus by submerged fermentation using corn husks as substrates. Int. Food Res. J. 22, 1831–1836.

Nigam, P.S., Pandey, A., (2009) Solid-state fermentation technology for bioconversion of biomass and agricultural residues. Biotechnol. Agro-Ind. Res. Util., 197, 221.

Nikodinovic, J., et al., (2008) The conversion of BTEX compounds by single and defined mixed cultures to medium-chain-length polyhydroxyalkanoate. Appl. Microbiol. Biotechnol., 80(4): p. 665–73.

P.F. Stanbury, A. Whiitaker, S.J.Hall, (1999) Principles of Fermentation Technology (Second ed.). Butterworth-Heinemann. ISBN 978–0750645010.

P.K.R. Kumar, B.K. Lonsane, (1987) Biotechnology and Bioengineering, Vol. 30, p. 267.

Pandey, A., Selvakumar, P., Soccol, C. R., Nigam, P., (1999) Solid state fermentation for the production of industrial enzymes. Curr. Sci. 77, 149–162.

Pleissner, D., et al., (2017) Direct production of lactic acid based on simultaneous saccharification and fermentation of mixed restaurant food waste. Journal of Cleaner Production, 143(Supplement C): p. 615–623.

R. Banerjee, G. Mukherjee, K.C. (2005) Patra, Bioresource Technology, Vol. 96, p. 949.

R.P. John, K.M. Nampoothiri, A. Pandey, (2006) Process Biochemistry, Vol. 41, p. 759.

Ramachandran, S., Patel, A. K., Nampoothiri, K. M., Francis, F. et al., (2004) Coconut oil cake-a potential raw material for the production of α-amylase. Bioresour. Technol. 93, 169–174.

Ramesh, M. V., Lonsane, B. K., (1987) A novel bacterial thermostable alpha amylase system produced under solid state fermentation. Biotechnol. Lett., 9, 501–504.

Rhu, D.H., et al., (2003) Polyhydroxyalkanoate (PHA) production from waste. Water Sci. Technol., 48(8): p. 221–8.

Rosales, E., Rodríguez Couto, S., Sanromán, A., (2002) New uses of food wastes: Application to laccase production by Trametes hirsuta. Biotechnol. Lett., 24, 701–704.

S. Bhargav, B.P. Panda, M. Ali, S. Javed, (2008) Chemical and Biochemical Engineering Quarterly, Vol. 22 (1), p. 49.

S. Miura, T. Arimura, I. Itoda, L. Dwiarti, J.B. Beng, C.H. Bin, M. Okabe, (2004) Journal of Bioscience and Bioengineering, Vol. 97, p. 153.

S.A. Shojaosadati, V. Babaeipour, (2002) Process Biochemistry, Vol. 37, p. 909, 2002.

Sanromán, M.A.L.a.M.A., (2006) Production of Food Aroma Compounds: Microbial and Enzymatic Methodologies. Food Technol. Biotechnol., 44(3): p. 335–353.

Schempp, F.M., et al., (2017) Microbial Cell Factories for the Production of Terpenoid Flavor and Fragrance Compounds.Journal of Agricultural and Food Chemistry.

Schmidt, C.G.; Gonçalves, L.M.; Prietto, L.; Hackbart, H.S.; Furlong, E.B. (2014)Antioxidant activity and enzyme inhibition of phenolic acids from fermented rice bran with fungus Rizhopus oryzae. Food Chem. 146, 371–377. [CrossRef] [PubMed]

Sellami, M., et al., (2015) Industrial wastewater as raw material for exopolysaccharide production by Rhizobium leguminosarum. Brazilian Journal of Microbiology, 46(2): p. 407–413.

Singh, V., Haque, S., Niwas, R., Srivastava, A., Pasupuleti, M., Tripathi, C.K.M., (2017) Strategies for fermentation medium optimization: an in-depth review. Front. Microbiol., 7, 2087.

Song, J.H., et al., (2008) Polyhydroxyalkanoate (PHA) production using waste vegetable oil by Pseudomonas sp. strain DR2. J. Microbiol Biotechnol, 18(8): p. 1408–15.

Stanbury, Peter F., Allan Whitaker, and Stephen J. Hall. (2016). Principles of fermentation technology, 3rd ed. London: Butterworth-Heinemann Press

Stredansky, M., Conti, E., Xanthan production by solid state fermentation. (1999) Process Biochem., 34, 581–587.

Subramaniyam, R., Vimala, R., (2012) Solid state and submerged fermentation for the production of bioactive substances: a comparative study. Int. J. Secur. Net., 3, 480.

T.D. Vrije, N. Antoine, R.M. Buitelaar, S. Bruckner, M. Dissevelt, A. Durand, M. Gerlagh, E.E. Jones, P. Lüth, J. Oostra, W.J. Ravensberg, R. Renaud, A. Rin - zema, F.J. Weber, J.M. Whipps, (2001) Applied Microbiology and Biotechnology, Vol. 56, p. 58.

Taskin, M., S. Erdal, and M. Genisel, (2011) Biomass and Exopolysaccharide Production by Morchella Esculenta in Submerged Culture Using the Extract from Waste Loquat (*Eriobotrya Japonica* L.) Kernels.Journal of Food Processing and Preservation, 35(5): p. 623–630.

Thomas, T.D., D.C. Ellwood, and V.M. Longyear, (1979) Change from homo- to heterolactic fermentation by Streptococcus lactis resulting from glucose limitation in anaerobic chemostat cultures. J. Bacteriol, 138(1): p. 109–17.

Tosungnoen, S., K. Chookietwattana, and S. Dararat, Lactic (2014) Acid Production from Repeated-Batch and Simultaneous Saccharification and Fermentation of Cassava Starch Wastewater by Amylolytic *Lactobacillus Plantarum* MSUL 702. APCBEE Procedia, 8(Supplement C): p. 204–209.

Tosungnoen, S., K. Chookietwattana, and S. Dararat, Lactic Acid Production from Repeated-Batch and Simultaneous Saccharification and Fermentation

Tunga, R., Shrivastava, B., Banerjee, R., (1558) Purification and characterization of a protease from solid state cultures of Aspergillus parasiticus. Process Biochem. 38, 1553– 1558.

Urbach, G., (1997) The flavour of milk and dairy products: II. Cheese: contribution of volatile compounds. International Journal of Dairy Technology, 50(3): p. 79–89.

Vandenberghe, L. P. S., Soccol, C. R., Pandey, A., Lebeault, J. M., (2000) Solid-state fermentation for the synthesis of citric acid by Aspergillus niger. Bioresour. Technol. 74, 175–178.

Venkateswar Reddy, M., et al., (2017) Polyhydroxyalkanoates (PHA) production from synthetic waste using Pseudomonas pseudoflava: PHA synthase enzyme activity analysis from P. pseudoflava and P. palleronii. Bioresource Technology, 234(Supplement C): p. 99–105.

Wee, Y., Kim, J., Ryu, H., (2006) Biotechnological Production of Lactic Acid and Its Recent Applications. Food Technology and Biotechnology, 44(2): p. 163–172, 2006.

X. Jian, C. Shouwen, Y. Ziniu, (2005) Process Biochemistry, Vol. 40, p. 3075, 2005.

Yang, S. S., Ling, M. Y., (1989) Tetracycline production with sweet potato residue by solid state fermentation. Biotechnol. Bioeng., 33, 1021–1028.

Yuliani, E., et al., (2011) Exopolysaccharide Production from Sweet PotatoShochu Distillery Wastewater by Lactobacillus Sakei CY1. Biotechnology & Biotechnological Equipment, 25(2): p. 2329–2333.

8

Biopharmaceutical Biochemical Properties Production and Applications

Swati Gupta and Shanthy Sundaram

Centre of Biotechnology, Nehru Science Centre, University of Allahabad Prayagraj. Uttar Pradesh-211002, India

Abstract

Fruits like orange (citrus sinensis), mango (mangifera indica), banana (musa paradisiaca), guava (psidium guajava) and pomegranate (punica granatum) contain numerous vitamins, minerals, fibre. These fruits also contain phytochemicals, antioxidant and bioactive compounds including alkaloids, flavonoids, tannins, saponin, terpenoids, which enhance the nutritional value and reduce the free radicals and oxidative stress in human body. These fruits also show the anti-cancer, anti-bacterial, anti-fungal, anti-inflammatory, anti-arthritic, anti-obesity, anti- diabetic, anti-proliferative, anti-hypertensive properties and are very effective against the pathogenic bacteria, which causes the harmful disease in human body. The chapter provides the production data of these fruits in different countries, their bioactive compounds and their utilization in the pharmaceutical, cosmetic and food industries.

Introduction

Orange (*Citrus sinensis*), mango (*Mangifera indica*), banana (*Musa paradisiacal)*, guava (*Psidium guajava*) and pomegranate (*Punica granatum*) are good source of the bioactive compound, vitamin, protein and minerals, which could very effective against the many harmful disease. These fruit are important part for the good diet. Secondary compound of the orange, guava, pomegranate, banana and mango fruits are also reduced the risk of several chronic diseases such as heart disease or cardiovascular disease, cancer and neurogenerative disorder (Negi and Jayaprakasha, 2001). These fruits also contained the flavonoids, which divided in to different group like favonols, flavones and flavanones, which prevent the harmful disease in human body such as diabetes, different type of cancer, obesity and inflammatory disease. Orange, mango, guava, banana and pomegranate fruits include the antioxidant compound, phenolic acid, pectin, carotenoids, alkaloids and other phytochemical compounds, which could reduced the oxidative stress in the cell. These fruits are contain the hesperidin and naringenin compound, which effective against the pathogenic disease and causes the serious illness in human body. Orange, banana, guava, mango and pomegranate fruits are show the antibacterial, antifungal, antiviral, antityphoid, anticancer, antituberculosis, antihepertensive and anti-inflammatory activities (Parmar and Kar, 2007). Orange, guava, mango, banana and pomegrante used in the pharmaceuitical industry, cosmetic industry and food industry, where many beneficially product such as jem, jellies, juice, and food supplement are made by these fruit. Different part of these fruits such as pulp, stem, root, seeds and peels are used in the therapeutic treatment and show the good medical properties (Parle and Chaturvedi, 2012). Orange, mango,banana, guava and pomegranatehave cultivated in every countries in the all over the world and total production of orange (80%), guava (57%), mango (42%), banana (149%) and promgrante (78%) fruits per year in the world.

Oange, mango, banana, guava, and pomegranate pulps and peels are rich in nutrients and contained bioactive and phytochemicals compounds (Parmar and Kar, 2007). Peels and pulps of these fruits can be used as drugs or food supplements and have the antibacterial, antiviral, antifungal activities and they can be used in same food industry, where it used as a food supplement (Huchina et al., 2014). Peels of orange, mango, banana, guava and pomegranate are rich source of flavanones and many polymethoxylated flavones (Zheng and Wang, 2001). Essential oils of these fruits showed the antimicrobial abilities in particularly interesting field and exhibited the applications in food and cosmetic industries (Parmar and Kar, 2007). It has also been used as an anti-diabetic, antimicrobial, antifungal, antioxidant, antibacterial, larvicidal, antiviral, uricosuric, anti-yeast, antihepatotoxic and antimutagenic agent (Lin et al., 1999). The antibacterial activity and the chemical constituents as well as the yield percentage of crude phytochemicals of these fruits has been quantified (Liew et al., 2018).

Antioxidants area natural and synthetic compounds, which able to scavenge free radicals and inhibit the oxidation processes in body's (Zoua et al., 2016). Phenolic acids form approximately one-third of the phenolic compounds in plant foods. They comprise hydroxycinnamic and hydroxybenzoic acid derivatives,which presented in the free and bound forms (Hayata et al., 2009). It is assumed that many antioxidative compound and phenolic compounds in these fruits are usually found in a covalently bound forms. Therefore reliable and practical methods for separation, identification and quantitative analysis of phenolic compounds are of considerable interest. Orange, mango, banana, guava, and pomegranate are contained active biological compounds include antioxidants, phenolic acids and flavonoids (Güzel and Akpınar, 2019). Phenolic acids was extracted by the different techniques including shaking, stirring and ultrasonic extraction (Kumar and Jayakuma, 2011).

Orange, mango, banana, guava and pomegranate are contained the medicinal properties. It is widely distributed in the tropical and subtropical regions. Orange, mango, banana, guava, and pomegranate include the natural antioxidants, pectin, enzymes, organic acids, essential oils, vitamin C, fibre, protein and many nutrients compounds andused as a traditional medicine such as stomach discomfort, skin inflammation, neuroprotectionand improving heart health (Güzel and Akpınar, 2019). Various antioxidants have been found in peels and pulps of these fruits and showed the free radical scavenging activities (Negi, 2012). It is exhibited the active phytochemicals in orange,mango,guava,banana and pomegranate fruits and play a main role in many diseases such as cancer, cardiovascular, neurodegenerative diseases and process of ageing (Parmar and Kar, 2007).

The world health organization (WHO) reported that about 85% of the population still depends upon herbal medicines for the treatment of various diseases due to easy availability, economic reasons and less side effects. Herbal remedies have formed the basis of traditional systems of medicine for ages and have formed the foundation of modern pharmacology. Phenolics are a bioactive compound, which present in high amount in peel and pulp of these fruits (Negi, 2012). Phenolic compounds of orange, mango, guava, banana and pomegranate fruits are provide the health benefits such as cardiovascular diseases, cancer, diabetes and obesity and prevent the infection of many disease.

Table 1: Botanical classification of Guava.

Kingdom	**Plantae**
Subkingdom	Tracheobionta
Division	Magnoliophyta
Class	Magnoliopsida
Sub-Class	Rosidae
Order	Myrtales
Family	Mytaceae
Genus	Psidium
Species	Psidium guajava

Table 2: Botanical classification of Banana.

Kingdom	**Plantae**
Subkingdom	Tracheobionta
Division	Spermatophyta
Class	Magnoliophyta
Sub-Class	Zindiberales
Order	Zingiberales
Family	Musaceae
Genus	Musa
Species	*Musa acuminata*

Table 3: Botanical classification of Orange.

Kingdom	**Plantae**
Division	Magnoliophyta
Class	Dicotyledons
Sub-Class	Sapindales
Order	Rosidae
Family	Rutaceae
Genus	Citrus
Species	*Sinensis*

Table 4: Botanical classification of Mango.

Kingdom	**Plantae**
Division	Magnoliophyta
Class	Magnoliopsida
Sub-Class	Rosidae
Order	Sapindales
Family	Anacardiaceae
Genus	MangiferaL.
Species	*Mangiferaindica L.*

Table 5: Botanical classification of promgrante.

Kingdom	**Plantae**
Division	Magnoliophyt
Class	Magnoliosida
Sub-Class	Rosidae
Order	Mytales
Family	Lythraceae
Genus	Punica
Species	*P. granatum*

Orange, mango, banana, guava, and pomegranate have been attributed with cooling, digestive and tonic properties and regulate the asthma, vomiting, cough, and fever. Similarly, pomegranate is used as digestive and cardiotonic agent and it highly effective in diarrhea and uterine disorders (Liew et al.,

2018). High amount of these fruits peel were showed the bioactive substances with antiperoxidative, antibacterial and anticancer activities (Gaoa et al., 2018). Recently, we observed the positive influence of the two peel extracts of orange and pomegranate on the levels of insulin, thyroid hormones and glucose in normoglycemic rats (Lin et al., 1999). Guava fruit is contains phytochemicals include polysaccharides, glycosides, steroids, flavanoids, tannins, saponins vitamins, essential oils, minerals, enzymes, proteins, alkaloids, saponins, sesquiterpenoid, alcohols,triterpenoid acids, antioxidants, vitamins, lutein, zeaxanthine and lycopene compounds (Wang et al., 2008).

Fig. 1: Different fruits

Orange, guava, banana, mango and pomegranate are contained phenolic compounds such asphenolic compounds phenolic acids, flavonoids, alkaloids, carotenoids and ascorbic acid (Negi, 2012). Phenolic compounds exhibited the various bioactivities like antimicrobial, antioxidant, anticancer, anti-inflammatory, antimutagenic and antiallergic properties (Singh et al., 2020). These fruits have the tangereti, nobilet compounds and they are showed the broad range of pharmacological application in food industries (Negi et al., 2008). Recently review study provides comprehensive information about the phenolic compounds, antimicrobial, antifungal, antioxidant activities and health benefits of these fruits (Milind and Dev, 2012).

Biochemical Properties of Different fruits

Flavonoids

Flavonoids are phenolic compounds,which have aromatic rings and bound with a 3-carbon bridge (C6–C3–C6 structure). These are major bioactive compound, which found in all of the fruits orange, mango, banana,guava,promgrante. Flavonoids include flavonols, flavanones,flavones and anthocyanins in different fruits (Zheng and Wang, 2001). Composition of flavonoid changes with the development and maturation of these fruits. Flavonoids are found naturally in all above mentioned fruits. Total flavonoid content change with the development of different fruits. The amount of different flavonoids (naringin, naringenin, hesperidin, hesperetin, rutin, nobiletin and tangeretin) were determined by high performance liquid chromatography (HPLC). They reported that the different immature fruits peel of these fruits contained hesperidin, naringin, nobiletin and tangeretin compounds. The main flavonoids identified in orange have hesperidin, naringin, tangeritin, rutin, flavonol (quercetin), and flavanones (naringenin and hesperitin). The total amount of flavonoids measured by HPLC–UVmethod, where the orange, mango, banana, guava, and pomegranate contained 16.45mg/g,8.24mg/g,22.43mg/g, 38.56mg/g and 22.04mg/g fruits peel weight respectively. It included flavanones (92.1%) and flavones (11.8%) as major and minor components (Kim et al., 2010). The main flavonoids are rutin (9.91%), naringin (5.23%), catechin (3.17%) and epicatechin (2.77%) (Zheng and Wang, 2001). Flavonoids have antioxidant enzyme and scavenging reactive oxygen activity, which also helped to improve the immunity in body's and decrease the peroxide formation (Y.C. Wang et al., 2008). Such as Hesperidin,Naringin, naringin, tangeritin, naringin, tangeritin have the antioxidant enzyme and scavenging reactive oxygen activity, which increases the activity of catalase (CAT), superoxide dismutase (SOD), glutathione peroxidase (GPx), paraoxonase (PON) and also decrease the tissue injury and peroxide formation. The content of naringin present in orange 12.92mg/100g (Negi et al., 2012). Flavonoids have DPPH scavenging activity and inhibit the oxidation of Cu^{2+} induced by in vitro and promote the pancreatic B cells (Parle, Chaturvedi, 2012). When the naringenin regulate the fatty acid oxidation than it inhibit the beta oxidation of fatty acids in the liver, therefore naringenins are more effective than other compounds. Quercetin and rutin are found in naringen, which presents in the different fruits (Parmar and Kar, 2008) (Table 2).

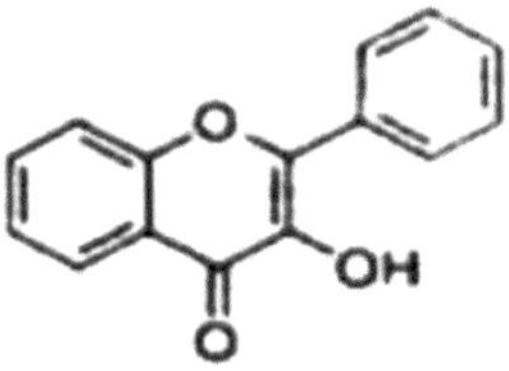

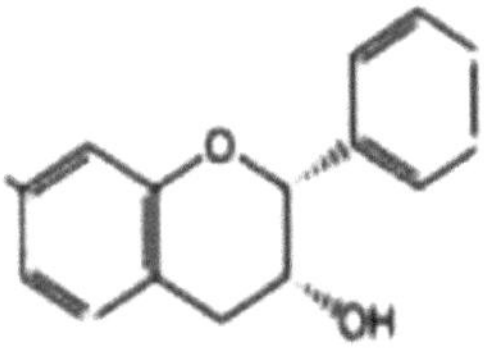

Flavonols

Flavanols

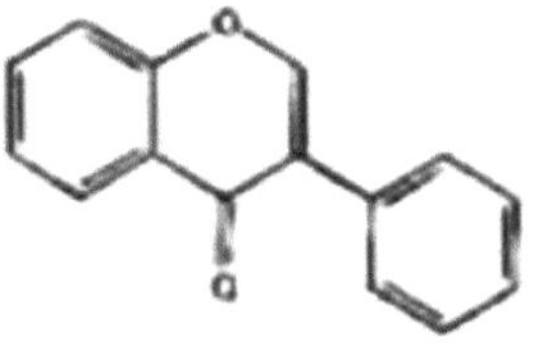

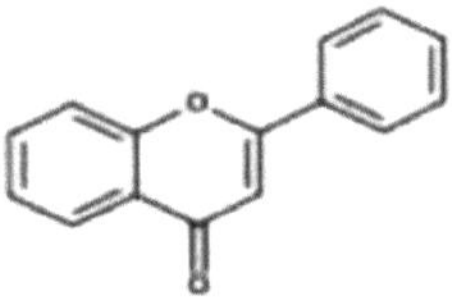

Isoflavones

Flavones

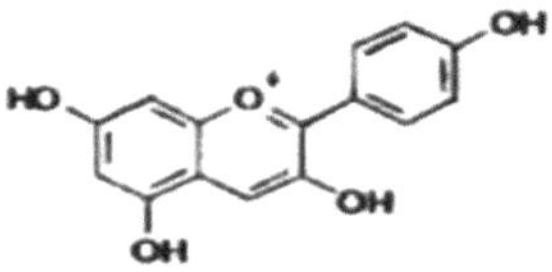

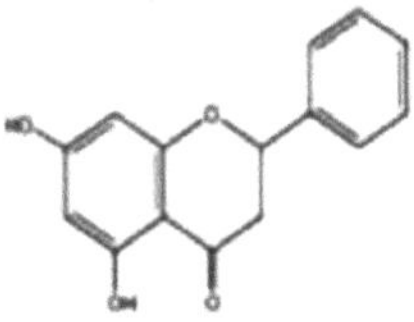

Anthocyanidins

Flavanones

Fig. 2: Different chemical structure of Flavonoids

Table 6: Phytochemical components Total phenolic,Total flavonoid,Total carotenoids and Ascorbic acid (mg/g) in fruit peels extract.

Scientific name	Local name	Total Phenolic	Total flavonoid	Total carotenoid	Ascorbic acid
Citrus Sinensis	Orange	86. 34	16.45	89.26	28.56
Mangifera indica	Mango	86.78	8.246	26.56	39.9
Musa paradisiacal	Banana	18.65	22.43	42.62	9.62
Psidium guajava L.	Guava	10.25	38.56	34.94	65.26
Punica granatum	Pomegranate	12.68	22.0.4	11.24	68.42

Flavanones

Orange, mango, banana, guava,and pomegranate fruits are contain more flavanone glycosides such as naringin, neohesperidin and neoeriocitrin (Kim et al., 2010).The composition of glycosylated flavanones varies in these fruits (Madhuri et al., 2014). The total neohesperidin content in methanol extracts of orange, mango, banana, guava,and pomegranate fruitspeel were reported as 22.45mg/g, 20.80mg/g, 11.34mg/g, 24.42mg/g, and 27.76 mg/g respectively (Singh et al., 2020). Orange peel contained a complex mixture of flavanones include hesperetin (aglycone), hesperidin and neohesperidin, while tangerine peel contained neohesperidin. narirutin, hesperidin, isosakuranetin and eriocitrin (Lin et al., 1999). The content of flavanone glycosides like neoeriocitrin, naringin, poncirin, dydimin and narirutin were reported as inflavedo extract in these fruits. Hesperidin is the most abundant flavonoid noted in dried peels of orange, which collected from USA and China with a level of 26.81 and 20.99 mg/g. The mean levels of hesperidin and naringinin of orangewas reported as 2.16 and 0.05 mg/g in dry peel. Total content of Naringin and hesperidin in orange, mango,banana, guava and pomegranate fruits are 34.16mg/g, 10.6mg/g, 40.4mg/g, 8.7mg/g, 78.6mg/g, and 0.48mg/g, 15.4mg/g, 1.56mg/g, 1.89mg/g, 29.94mg/g (DW) major flavanones compounds, reported in these fruits peel (Kim et al., 2006).

Table 7: Flavanones(mg/g) in different fruits

Scientific name	Local name	Naringin	Hesperidin	Neohesperidin
Citrus Sinensis	Orange	34.16	0.48	22.45
Mangifera indica	Mango	10.6	15.4	20.80
Musa paradisiacal	Banana	8.72	1.56	11.34
Psidium guajava L.	Guava	40.4	1.89	24.42
Punica granatum	Pomegranate	78.6	29.94	27.76

Flavones

Flavones represent the most diverse source of flavonoids with nobiletin, tangeritin and diosminas the main PMF's (Afroja et al., 2017). The contents of three PMF's heptamethoxyflavone, tangeretin and nobiletin in orange peel extract were 55.04mg/g,20.34mg/g and10.41mg/g, respectively. Methanolic peel extract of ponkan contained nobiletin, tangeretin, and sinensetin at the level of 5.89mg/g,6.41mg/g and0.47mg/g,respectively while tonkan contained these compounds at a level of 6.93, 1.43 and 0.64 mg/g, respectively. Nobiletin (92.5 mg/kg FW) and tangeretin (55.9 mg/kg FW) were the major flavones identified and quantified by HPLC–UV method in pulp and peel of these fruits. The other flavo compounds are identified in pulp and peel of these fruits were 3,3′,4′, 5,6,7,8-heptamethoxyflavone, sinensetin, hexamethoxyflavones and isosinensetin. Total content of nobiletin and tangeretin compound were prepared from heated peel powder 150°C for 160 minuts and have the 40.1mg/g, 36.7mg/g, 39.2mg/g,40.1mg/g and 32.4mg/g) extracts in orange, guava, pomegranate, mango and banana fruits, whereas non-heated powder content was (29.5mg/g, 26.7mg/g, 24.2mg/g, 28,9mg/g and 25.9mg/g) extract, respectively (Singh et al., 2020). Flavones were more efficaciously extracted by microwave-assisted and soxhlet extraction method from peel of these fruits, respectively (Negi et al., 2012). It is less time-consuming extraction method and has higher extraction efficiency as compared to other extraction techniques.

Flavonols

Organic extract of orange, mango, banana, guava and pomegranate fruits have the flavonols such as limocitrol, isolimocitrol, limocitrin and rutin. The amounts of flavonols (quercetin) was reported in orange, guava, pomegranate, banana and mango are 15.3 mg/g, 67.3 mg/g,7.45mg/g,13.2 mg/g and 58.4mg/g,respectively. The content of rutin present in flavedo extracts of orange, mango,banana, guava and pomegranate was reported as 42.13mg/g, 33.13mg/g, 13.09mg/g, 10.62mg/g and 8.16mg/g DW respectively (Zhou, 2016).The level of kaempferol and rutin was reported in fresh fruits of orange, mango, guava, banana and promgrenate 4.36mg/g,52.5mg/g,34.6mg/g, 42.6 mg/g, 98.5mg/g and 18.45mg/g, 0.98 mg/g,6.42mg/g,3.45mg/g and 19.5mg/g and content was increased to 6.63 and 5.31 mg/g, respectively after drying of orange, mango, banana, guava, and pomegranate fruit at 100°C. The amount of rutin in *citrus sinensis* peels extract obtained by using solvent extraction methods was18.45μg/g (Li et al., 2006).

Table 8: Flavonol contents (mg/g) in Different Fruits.

Scientific name	Local name	Rutin	Quercetin	Kaempferol
Citrus Sinensis	orange	18.45	15.3	4.36
Mangifera indica	Mango	0.98	67.3	52.5
Musa paradisiacal	Banana	6.42	7.45	34.6
Psidium guajava L.	Guava	3.45	13.2	42.6
Punica granatum	Pomegranate	19.5	58.4	98.5

Carotenoids

Carotenoids are tetraterpenoids, antioxidant activities and eliminate the harmful free radicals in the body's. Carotenoids protect the immune cell membrane lipids by oxidative damage, they provide communication signals between cells and receptors on the cell membrane and enhance the human immunity (Zhou, 2016). The total content of carotenoids in orange, mango, banana, guava and pomegranate fruits peel are 8926.mg/g, 26.56mg/g, 42.62mg/g, 34.94mg/g and 11.24mg/g,respectively.

Limonoids

Limonoids are oxygenated group, which have tetracyclic secondary metabolite. Limonoidaglycones and limonoid glycosides are present in these fruits (Zhou, 2016). Different limonoids contain variable antioxidant activity. For example, the free radical scavenging activity of four limonin glycosides including limonin 17-b-D-glucoside (LG), obacunone 17-b-Dglucoside body (OG), millington acid 17-b-D-glucoside (NAG) and deacetylationmillington acid 17-b-D-glucoside (DNAG) were evaluated. It was found that all of them have antioxidant activities, where NAG is the strongest and LG is the weakest. The content of limonoids in these fruits ranges from 0 to 97.48mg/100 g (Zoua et al., 2016).

Coumarins

Coumarins are rich source of phenolic compounds, which present in these fruits. They are derived from phenylalanine metabolism pathway, which synthesis the furanocoumarin (psoralin). Coumarins have phenolic hydroxyl groups and shown the strong antioxidant activities (Li et al., 2006). The content of coumarins in orange, mango, banana, guava, and pomegranate are 12.3mg/g,25.5mg/g,10.4mg/g,0.89mg/g and 19.35mg/g, respectively (Zhou, 2016).

Pectin

Pectin is the major compound of the cell wall in plants. It is a polysaccharide chemical, which consist the linear chain of galacturonic acid. It reduced the lipid level in blood and exhibited the antioxidant activities in kidney, where the toxicity of kidney induced by octylphenol. When pectin show the suppressive effect than it enhance the endogenous antioxidant enzymes andharmfull free radicals (Table 1). Usually high pectin contentspresent in orange, mango, banana, guava and pomegranate36.0± 1.46mg/g db to 86.4 ±3.36 mg/g, respectively (Zoua et al., 2016).

Ascorbic Acid

Vitamin C (ascorbic acid) is a water soluble material and present in a rich amount in citrus pulp and peel of different fruits. Contents of ascorbic acid are investigated pulps and peels of orange, guava, mango, banana and pomegranate fruits. Ascorbic acid was determined by 2,6-dichloroindophenol titrimetric method. Their corresponding concentrations of fresh peels were found 28.56mg/g, 39.9mg/g, 9.62mg/g, 65.26mg/g and 68.42 mg/100g dry weight basis (Table 2). When drying the orange peel and pulp by air oven, then it reduces the concentration of the ascorbic acid and it have half content from the original values (control). Vitamin C present in a higher amount in orange peels compared to other fruit peels and pulps (Zoua et al., 2016).

Phenolic acids

Phenolic acids are rich sources of these fruitsand have the free radical scavenging activity. Phenolic acids exhibited antioxidant properties in orange, mango, banana, guavaandpomegranatefruits. Phenolic acids have hydroxyl groups and affect the ortho-substitution on a benzene ring (Singh et al., 2020). The total content of phenolic acids different in each fruit is given below (Table 1) (Wang et al, 2008).

Table 9: Phenolic acid (mg/g)in different fruit

Scientific name	Local name	Caffeic acid	Ferulic acid	Sinapic acid	p-Coumaric acid
Citrus Sinensis	orange	4.59	10.2	40.2	12.3
Mangiferaindica	Mango	2.18	14.5	0.8	25.5
Musa paradisiaca	Banana	5.67	6.52	1.6	10.4
Psidium guajava L.	Guava	2.56	1.6	80.5	0.89
Punica granatum	Pomegranate	25.7	18.5	26.5 19.3	

Mineral elements

Mineral elements are chemical compounds. (Zhou, 2016). Orange, mango, banana, guava and pomegranate fruits include calcium (Ca), phosphorus (P), magnesium (Mg), potassium (K), sulfur (S), sodium (Na), iron (Fe), manganese (Mn), nickel (Ni), boron (B), silicon (Si), copper (Cu), zinc (Zn), molybdenum (Mo), selenium (Se), cobalt (Co), chromium (Cr), germanium (Ge) and arsenic (As) (Kaviya et al.,2011). Some elements of Mn, Fe, Cu, Zn and Se show the antioxidant activities in organisms. Mn,Cu and other minerals can destroy the free radicals and protect the tissues against the oxidative damage (Zhou, 2016). The content of calcium and phosphorus in the extraction of orange,mango,banana, guava and pomegranate are 0.31mg/100g, 0.45mg/100g, 0.50mg/100 g, 0.48mg/100g and 0.47mg/100g, respectively (Zheng and Wang, 2001).

Production of different fruits

The global average production of orange (4.80%), guava (4.56%), mango (16.04%), banana (89%) and pomegranate (12.88%) in a year all over the world. These different fruits are widely cultivated in tropical and subtropical climates in more than 190 countries including China, India, Brazil, Indonesia, Philippines and Egypt. Brazil produced a large amount of the different fruits such as orange (43%), mango (5%), banana (11%), guava (4%) and pomegranate (69%) followed by China producing orange (22%), mango (16%), banana (21%), guava (17%) and pomegranate (2%). India is the third in this sequence, which produced orange (19%), mango (68%), banana (46%), guava (68%) and pomegranate (7%). Fourth biggest country in production of these fruits is Indonesia, which produced orange (5%), mango (8%), banana (11%), guava (5%) and pomegranate (12%) followed by Philippines produced orange (3%), mango (3%), banana (9%), guava (3%) and pomegranate (6%). At the end Egypt produced orange (8%), mango (4%), banana (2%), guava (3%) and pomegranate (4%) in the world. Citrus fruits cultivated in new areas, where other different varieties of the fruits are not damage in a harsh condition of any season but old areas is inhibited the total production of the different fruits of citrus due to urbanization fastly. Diseases and harsh condition of the season are the major factors, which effected the cultivation of these fruits. As the costs of land, labour, and fuel rise, new technologies will also helped to increase the productivity of the citrus fruits in the world, and developed to increase the productivity of land for fruits to be competitive in the world market.

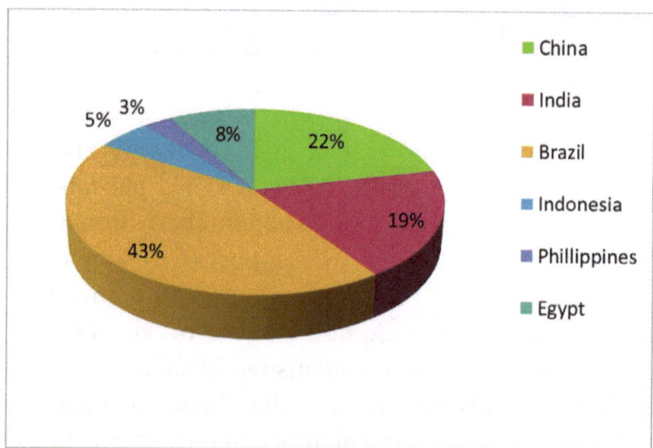

Orange fruits production by countries.

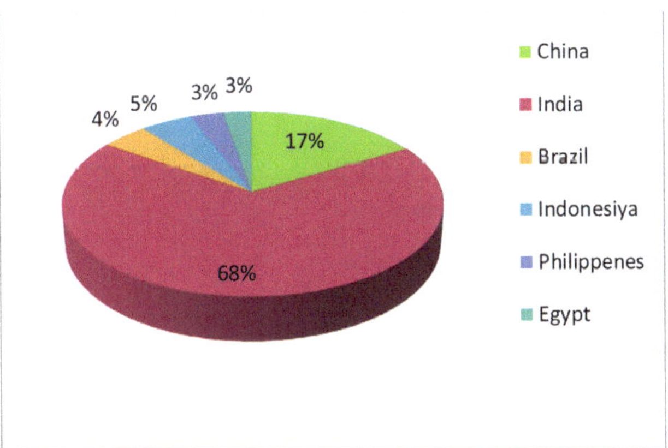

Guava fruits production by countries.

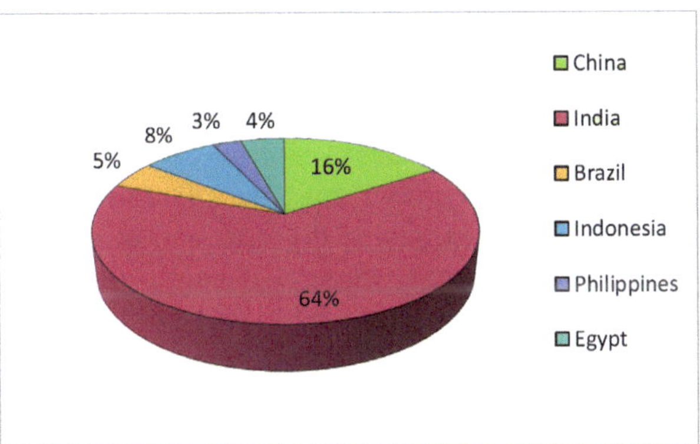

Mango fruits production by countries.

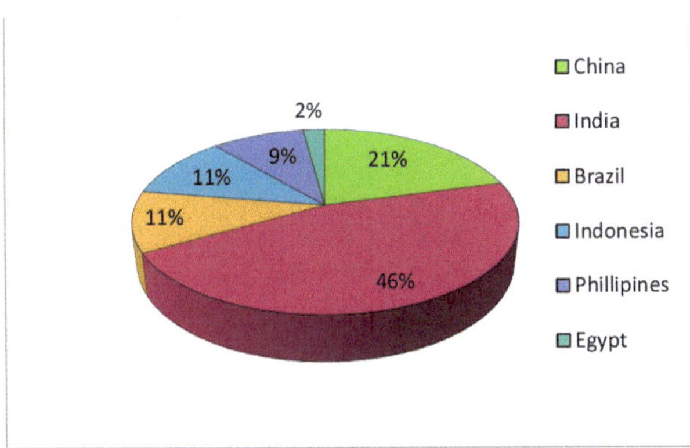

Banana fruits production by countries.

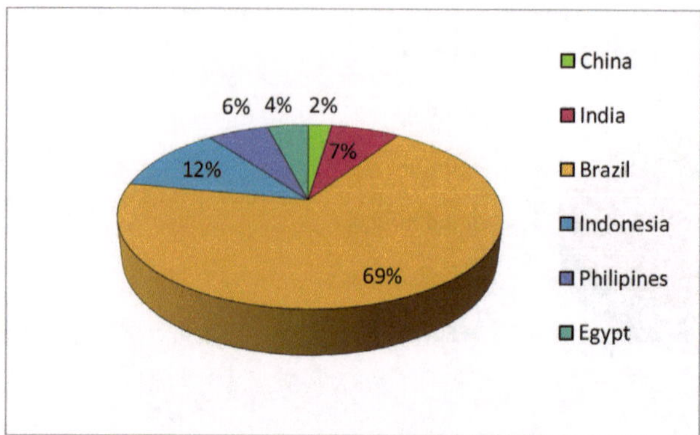

Pomegranate fruits production by countries.

Fig. 3: Production data of fruits in different countries (Tons)

The fruits such as orange, mango, banana, guava, and pomegranate have the vitamin, minerals, starch, soluble sugar, cellulose and hemicelluloses fibre, pectin, lignin, dietary fibre, fat and protein and showed the biological activity including antiviral, anti-obesity, anti-inflammatory, antibacterial, antifungal, anti-diabetic, anti-proliferative and anti-typhoid activities. Different fruits showed the medicinal properties, which minimize the infection of different disease and enhance the immunity. Orange, mango, banana, guava and pomegranate contained the many secondary compounds such as flavonoids, alkaloids, tannins, terpenoids, saponin and carotenoids, which showed the physiological and pharmaceutical properties. It have the great potential to inhibit the harmful disease in the human body's and provide the protection against the several disease.

Applications of Different Fruits

Antioxidant and Antimicrobial activities of different fruits

Pomegranate (*Punica granatum*)

Pomegranate fruit has contained many antioxidant compounds such as alkaloids, saponins, tannins, flavonoids and phenolic compounds, which have healing characteristics.Various extracts of pomegranate fruit peels were prepared by different extraction methods. It showed antimicrobial activity against the many pathogenic bacteria by the in-vitro agar diffusion method. It was found that 70% methanolic extract and other organic extracts of peels and pulp of pomegranate was a potent inhibitor for *Bacillus subtillis, Salmonella typhi, Staphyllococcus aureus* and *Escherichia coli* bacteria. Pomegranate pulp and peels have phenolics and flavonoids compounds, which are active inhibitors against pathogens (Afroja et al., 2017).

Mango *(Mangifera indica)*

Mango is a seasonal fruit, which contained antioxidant and phenolic compounds. It has antioxidant properties and destroys harmful free radicals in the tissue.The antioxidant activity of peels of mango in the organic extract was determined by using different antioxidant methods such as reducing power activity and DPPH free radical scavenging activity.The results revealed that distinct organic extracts of mango peels comprised anthocyanins and carotenoids, both of which exhibits high DPPH and reducing power activity (Parashar et al., 2014).

Orange (*Citrus sinensis*)

Citrus is a medicinal plant that belongs to the Rutaceae family, Citrus fruits are good sources of vitamin C and flavonoids. Orange peels were found to have high antioxidant activity in each extract. Ethanolic and methanolic extract of orange fruits have the highest values of total phenolic content, and total flavonoid content (Kim et al., 2006).It also showsthe antifungal and antimicrobial activity (Zoua et al., 2016).

Table 10: Estimation of Antioxidant Total antioxidant capacity (%), Total reducing power (%)] in different fruit extracts.

Parameter	Total antioxidant capacity	Total reducing power
Banana	91.23 ± 0.137	93.48 ± 0.812
Guava	32.93 ± 0.663	29.18 ± 0.391
Mango	59.38 ± 0.723	38.71 ± 0.417
Orange	85.98 ± 0.993	81.24 ± 0.729
Pomegranate	87.39 ± 0.780	90.29 ± 0.562

Banana (*Musa paradisiaca*)

Banana is a tropical plantand one of the most popular fruit worldwide. Banana peels are rich in dietary fibre, proteins, essential amino acids, polyunsaturated fatty acids and potassium. The antioxidant activities of the extracts obtained from fresh green and yellow banana fruits were evaluated by using the thiocyanate method, ß-carotene bleaching method and 1,1diphenyl-2-picrylhydrazyl free radical elimination method. The fresh green and yellow banana fruits were treated with 70% acetone, which was then partitioned with chloroform and ethyl acetate (Negi., 2012). The antimicrobial activities of the extracts and isolated components were evaluated using paper disc methods for minimum inhibition concentration. The water soluble and ethylacetate fractions of green peel displayed high antimicrobial and antioxidant activity. Antioxidant activity of water extracts was more pronounced and comparable to those of synthetic antioxidants such as butylatedhydroxyanisole and butylatedhydroxytoluene. Among all isolated components succinic acid, ß-sitosterol, palmitic acid, malic acid, 12-hydroxystearic acid, glycoside, d-malic and 12hydroxystrearic acid were the most active against all the gram-negative and gram-positive bacterial species tested (Wang et al, 2008).

Guava (*Psidium guajava L.*)

Guava is a medical plant with several medicinal properties, including anticancer, antibacterial, antifungal, antityphoid, and anti-inflammatory activities (Negi et al., 2008).The antimicrobial activity of guava fruit has been determined by the minimum inhibitory concentration by using different solvents. It is effective against many pathogenic bacteria such as *Pseudomonas aeruginosa, Salmonella typhimurium, Micrococcus aureus*. The peels and pulp of the guava fruit also revealed antioxidant activity (Zoua et al., 2016).

Anti-inflammation

Inflammation isa normal biological process,which causes tissue injury, microbial pathogen infection and chemical irritation. Overproduction of inflammatory mediators such as free radicals and cytokines causes deterioration of tissue function. Prostaglandins including prostaglandin E2 (PGE2) produced by the inducible enzyme Cyclo-oxygenase (COX). PMFs exert anti-inflammatory potential through inhibiting gene expression and enzymes involved in inflammation. Nobiletin and tangeretin can be considered as anti-inflammatory and nutraceuticals. They suppressed LPS and immunoglobulin E mediated activation of human intestinal mast cells by producing critical pro-inflammatory mediators and signaling pathways (Parle and Chaturvedi, 2012). Nobiletin hinders Ca^{2+} entry into cells and suppresses or regulates degranulation and phosphorylation of enzymes such as PI3K and phospholipase Cγ1 in signaling pathways, which is related to degranulation and stimulated by antigen.

Anti-diabetes

Normal insulin production is reduced the pancreas, to cause Type 2 diabetes. Tangeretin stimulated glucose uptake in cell culture and regulate the AMP-activated protein kinase (AMPK) signaling pathway (Parmar, Kar.,2007). Tangeretin decreases plasma glucose and glycosylated hemoglobin levels in daily oral administration while increasing insulin and hemoglobin levels, which are essential enzymes in carbohydrate metabolismssuch as hexokinase, pyruvate kinase, lactate dehydrogenase, glucose-6-phosphatase, fructose-1, 6-bisphosphatase, glucose-6-phosphate, dehydrogenase, glycogen synthase and glycogen phosphorylase in the liver of diabetic rats were reverted to near-normal levels. The effect produced by tangeretinequivalent to the antidiabetic drug glibenclamide treatment group. Nobiletin was found to improve hyperglycemia and insulin resistance in diabetic ob/ob mice and improved thelevel of plasma glucose (Zheng, 2001).

Anti-obesity

Obesity is closely associated with adipocyte dysfunction.It is a metabolic disease like hypertension, hyperlipidemia and arteriosclerosis. Several PMFs showed anti-obesity activity. Different extracts of fruits contained several PMFs, which exhibited the anti-obesity activity on HFD induced obese C57BL/6 mice and mature 3T3-L1 adipocytes. PMFsextract with dietary supplements significantly decreased the body weight gain, adipose tissue weight and triglyceride in HFD mice. Moreover, it significantly decreased the accumulation of fatty droplets in the liver, and HFD-induced decrease the phosphorylation levels of AMPK and ACC, which are related to fatty acid β-oxidation (Parle and Chaturvedi, 2012).

Anti-cancer

Cancer has long been a primary cause of death in the world. Mostly, the oncogenesis process comprises three stages: initiation, promotion and malignant progression. PMFs of different fruit exhibited anti-cancer activity in each stage (Zhao et al., 2018).

Anti-mutagenic action

Mutations in DNA repair or cell cycle checkpoint genes result genomics instability and in turn may give rise to tumor genesis. Tangeretin from various fruits suppressed cancer cells and reduced DNA double-strand break repair when the SGS1 gene was deficient (Parle and Chaturvedi, 2012). It increases the DNA damage in yeast cells and gives the responses for the repair of DNA double-strand breaks and causes cell death.Therefore, flavonoids, phenolic acid, and other compounds aided in the prevention of cancer cells generated by mutations (Zheng and Wang, 2001).

Anti-angiogenesis action

Angiogenesis provides a suitable condition for tumor growth. Tumor cells take nutrition and oxygen from the blood vessels. In recent studies, tumor growth and angiogenesis were inhibited bynobiletin in ovarian cancers. Secretion of nobiletininhibits the angiogenesis mediatorsHIF-1α, NF-κB and vascular epithelial growth factor (Singh et al., 2020).

Regulation of metabolic disorder

The prevalence of metabolic syndrome has increased the risk of metabolic and cardiovascular diseases around the world. Metabolic syndrome is characterized by various metabolic abnormalities, such as central obesity, hyperlipidemia, hyperglycemia, and hypertension. Aberrant lipogenesis, inflammation is elevated and reduced fatty acid oxidation. Other metabolic pathways are implicated with the occurrence of metabolic syndrome (Parle and Chaturvedi, 2012). PMFs have an obvious effect on the control of lipid and glucose metabolism, which helps to alleviate metabolic disorders.

Anti-atherosclerosis

The Progression of atherosclerosis is complicated and stimulates cholesterol and free radicals. It attaches to adhesion molecules and promotes the growth of vascular endothelial cells, resulting in atheroma formation. Flavonoids of different pulp and peels of orange,mango,banana,guava and pomegranate fruits prevent atherosclerosis (Kim et al., 2006).

Neuroprotective effects

Neurodegenerative diseases, such as Alzheimer's disease (AD), Parkinson's disease have become a huge burden in the aging society.Degenerations of these neural cells result from many factors, including oxidative stress, neuro-inflammatory and the alteration of the signal path. Increasing evidence has shown that PMFs of different fruits exhibited the characteristics of neuroprotective molecules via the intervention of neurodegeneration and promotion of brain functions.Oxidative stress is considered the major cause of neurodegenerative disorders and PMFs could exert neural cells protection against oxidative stress. Early studies showed that pre-treatment with the PMFs extract of tangerine peelcould attenuate the dopaminergic neurotoxin in-vivo(Y.-C. Wang et al.,2008).Tangerine and nobiletin in ethanolic extract of Citrus *sunki peel*, was found to markedly increase HT-22 neurons viability against H_2O_2-induced cell death in a dose-dependent manner, which is associated with their inhibitory effect on H_2O_2-induced mitochondrial Ca^{2+} overload, therefore, avoid neuronal apoptosis (Negi, 2012).

Anti-typhoid activity

Typhoid fever is caused by *Salmonella typhi* and is a severe health concern in humans. The fruits of orange, mango, guava, banana, and pomegranate had antityphoid activity, which included phenolic, flavonoids, and other phytochemical compounds (Parle and Chaturvedi, 2012).

Anti-fungal activity

Essential oil of orange, mango, banana, guava and pomegranate fruits exhibited the anti-fungus activity in different fungus (Kim et al., 2006).

Conclusion

The fruits such as orange (*Citrus* sinensis), mango (*Mangifera indica*), banana (*Musa paradisiaca)*, guava (*Psidium guajava*) and pomegranate (*Punica granatum*) are the most popular fruits in the world. These fruits contain the phytochemical compounds including alkaloids, flavonoids, tannins, saponin, terpenoids etc., which enhance the nutritional value in the human body. These fruits contain the many bioactive compounds, which are very effective against many pathogenic bacteria, which cause the harmful diseases in the human body. Different parts like pulp, root, shoot, leaf, bark, seed, and flower of these fruits are very useful for the therapeutic treatment. These fruits also show the anti-cancer, anti-bacterial, anti-fungal, anti-inflammatory, anti-arthritic, anti-obesity, anti-diabetic, anti-proliferative, anti-hypertensive properties. Different fruits contain the different antioxidant compounds, which reduce the free radicals in the cell and decrease the oxidative stress in the human body. These fruits also have vitamins, minerals, fibre, and other secondary compounds, which show the medical properties against the harmful diseases. These fruits are cultivated in almost every country and used in the pharmaceutical, cosmetic, and food industries, where many products such as jellies, jams, juice, and food supplement are formed by the these fruits.

References

Al-Snafi, A.E.(2016)." Nutritional value and pharmacological importance of citrus species grown in Iraq IOSR."Journal of Pharmacy, 6(8),76-108.

Ames, B.N., Shigenaga, M.K., & Hagen, T.M. (1993). "Oxidants, antioxidants and the degenerative diseases of aging." Proceedings of the National Academy of Sciences of the United States of America, 90,79157922,19.

Amutha, R., Kavusik,T., & Sudha. A. (2017)."Analysis of Bioactive Compound in Citrus Fruit Peels." International Journal of Scientific Research and Review,2279-543X.

Anwar,F., Naseer,R., Bhanger,M.I.,Ashraf,S.,Talpur, F.N., & Aladededune, F.A.(2008). "Physicochemical characteristics of citrus seeds and oils from Pakistan."J. Am. Oil Chem. Soc., 85, 321, 330-24.

Babbar, N., Oberoi, Singh.H., Uppal, S.D., Rambhau., & Patil, T. (2011). "Total phenolic content and antioxidant capacity of extracts obtained from six important fruit residues."Food Research International, 44,391-396.

Bocco, A., Cuvelier, M.E., Richard, H., &Berset, C. (1998). "Antioxidant activity and phenolic composition of Citrus peel and seed extracts."J. Agric. Food Chem. 46(6), 2123,2129-25.

Chowdhury,S.A., Sohrab, M.H., Datta, B.K.,&Hasan, C.M.(2008). "Chemical and Antioxidant Studies of Citrusmacroptera, Bangladesh."Journal of Scientific and Industrial Research, 43(4), 449-454.

Dhanavade, J.M., Jalkute,B.C., Ghosh,S.J., &Sonawane,D.K.(2012). "Study Antimicrobial Activity of Lemon (*Citrus lemon* L.) Peel ExtractBritish."Journal of Pharmacology and Toxicology(3): 119-122.

Dorta, E., González, M., Lobo, M. G., &Laich,F. (2015). "Antifungal activity of mango peel and seed extracts against clinically pathogenic and food spoilage yeasts."Taylor&Francis, 1478-6427.

Espiard, E. 2002."Introduction à la transformation industrielle des fruits." (Ed) TEC &DOC. France,59, 265-31.

Etebu, E., &Nwauzoma, A.B.(2014)."A Review on Sweet Orange (*Citrus sinensisl* L Osbeck)."Health, Diseases And Management, 2(2), 33-70.

Goulas,V.,&Manganaris,G.A. (2012). "Exploring the phytochemicalcontent and the antioxidant potential fruits grows in Cyprus."Food Chemistry, 131,39-47.

Griffn,G.S.,& Markham, L.J.(2000)."An Agar Dilution Method for the Determination of the Minimum Inhibitory Concentration of Essential Oils."J. Essent. Oil Res,12, 249-255.

Gupta,S., Rahman,M.A., &Sundaram, S.(2021). "Citrus Fruit as a Potential source of phytochemical, antioxidant and pharmacological ingredients.".Journal of Science and Healthcare Exploration (JSHE),2581-8473.

Hang, T.V., Christopher, J. S., &Quan, V. V.(2018)."Phenolic compounds within banana peelandtheir potential uses: A review."Journal of Functional Foods, 40, 238–248.

Hayata,K.,Hussainb, S., Abbasa,S., Farooqa, U., Dinga, B.,Xiaa, S., Jiaa, C., Zhanga, X., &Xiaa, W.(2009). "Optimized microwave-assisted extraction of phenolic acids from citrus mandarin peels and evaluation of antioxidant activity in vitro." Sep Purif Technol,70,63–70.

Ismaill, H., Chan, K.W., Mariod, A.A., Ismail, M. (2010). "Phenolic Content and antioxidant activity of cantaloupe(cucumismelo)methanolicextracts."Food Chemistry,119,643-647.

Ke, Y., J. XQ, Z., Ge, H., G., Ma. Y., M.,& Chang, Z. H. (2015). "Low field induced giant anisotropic magneto caloric effect in DyFeO3 single crystal."ChinesePhys. B., 24,037501.

Kim, H.K., Tsao, R., Yang, R., & Cui, W.S. (2006). "Phenolic acid profiles and antioxidant activities of wheat and bran extracts and the effect of hydrolysis condition." Phenoli acid profiles and antioxidant activities of wheat bran extracts and the effect of hydrolysis conditions. Food Chemistry, 95, 466–473,

Kim,H.,Moon,Y.J.,Kim,H., Lee,S-D.,Choj, M.,&Choi,K.H.C.(2010)."Antioxidant and antiproliferativeactivitiesof mango (*Mangiferaindica* L.) flesh and peel."Food Chemistry, 121,4.

Kulkarni, T.R., Mateenuddin, M., Bodhankar, S.L.,&Saharabudhe, R.A. (2012)."Reversible anti- fertility effect of lemon seeds (*Citrus limonum*) in male albino rats."International Journal of Research in Pharmaceutical and Biomedical Sciences,3(2), 545,550-78.

Lagha,B., &Samir,M.K. (2013). "Phenolic content and antioxidant activity of orange varieties (*Citrus sinensis* L. and *Citrus aurantium* L.) Cultivated in Algeria:Peel and leaves". Industrial Crops and Products, 50,723-730.

Lahlou,M. (2003). "Biological and Pharmacological Approaches to the Screening and Evaluation ofNaturalProducts". Les Approaches Pour Le screening Etl'evaluation Des Produits Naturals, 58 (6),535-539.

Liew, S. S., Ho, W. Y., Yeap, S. K., &Sharifudin, S. A. B. (2018). "Phytochemical composition and in vitro antioxidant activities of Citrus sinensis peel extracts."PeerJ,5331.

Lin, J.,Opoku, A. R., Keller, M. G., Hutchings, A. D., Terblanche, S. E., A. Jäger, A. K..,&Staden, J. V.(1999). "Preliminary screening of some traditional zulu medicinal plants for anti-inflammatory and anti-microbial."Journal of Ethnopharmacology, 68, 267–274.

Madhuri S., Ashwini U. H., Srilakshmi N.S., &Kekuda T.R. P. (2014)."Antimicrobial Activity of Citrus sinensis and Citrus aurantiumpeelextracts."Journal of Pharmaceutical and Scientific Innovation,10.7897/2277-4572.034174.

Middleton, E., &Kandaswami, C. (1994). "Potential health-promoting properties of Citrus flavonoids."FoodTechnol., 115,119-21.

Montanari, A., Chen, J., &Widmer, W. (1998). "Citrus flavonoids: a review of past biological activity against disease. In: Manthey, J.A., Buslig, B.S. (Eds.), Flavonoids in the Living System." Plenum Press, New York,103,113- 22.

Moo-Huchina, V. M., Moo-Huchinb, M. I., Estrada-Leóna, Luis Cuevas, L.,Gloryc, Estrada-Mota, I. A., Ortiz-Vázquezc, E., Betancur-Anconad, D.,&Sauri-Duchc. E. (2015). "Antioxidant compounds, antioxidant activity and phenolic content in peel from three tropical fruits from Yucatan, Mexico."FoodChem,, 166,17–22.

Nath, A. Tiwari, K.P.,Rai,K,A., &Sundaram, S.(2019). "Evaluation of carbon capture in competent microalgal consortium for enhanced biomass, lipid, and carbohydrate production"3 Biotech, 9, 11-379.

Negi,P.S. (2012). "Plant extracts for the control of bacterial growth: efficacy, stability and safety issues for food application."Int J. Food Microbiol, 156,7–17.

Negi,P.S., &Jayaprakasha,G. K. (2001). "Antibacterial activity of grapefruit (citrus paradisi) peel extracts."Eur. Food Resour.Technol., 213,484–487.

Parashar,S., Sharma, H., &Garg, M. (2014). "Antimicrobial and Antioxidant activities of fruitsand vegetable peels:" A review Journal of Pharmacognosy and Phytochemistry, 3 (1), 160-164.

Parle,M., &Chaturvedi,D.(2012). "Orange: Range of Benefits."International Research Journal Of Pharmacy, 3 (7), 2230–8407,

Parmar, S.H.,&Kar A.(2007)." Antidiabetic potential ofCitrussinensisis and Punicagranatum peel extracts in alloxan treated male mice."BioFactors, 31,17–24.

Parmar,H.S.,&Kar,A.(2008)."Antiperoxidative, antithyroidal, antihyperglycemic and cardioprotective role of citrus sinensis peel extract in male mice." Phytotherapy Research: An International Journal Devoted to Pharmacological and Toxicological Evaluation of Natural Product Derivatves, 22 (6) 791,795.

R′ıos,J.L., &Recio, M.C. (2005). "Medicinal plants and antimicrobial activity." Journal ofEnthopharmacology, 100,80-84.

Rajendran, P., Nandakumar, N., Rengarajan, T.,Palaniswami, R., Gnanadhas, N.E., Lakshminarasaiah, U, Gopas, J., &Nishigaki,I.(2014). "Antioxidants and human diseases."ClinChimActa, 436,332–347.

Samman, S., Wall, P.M.L.,& Cook, N.C. (1996). "Flavonoids and coronary heart disease: dietary perspectives. In: Manthey, J.A., Buslig, B.S. (Eds.), Flavonoids in the Living System." Plenum Press, New York, 469,481-23.

Singh,B., Singh, P J., Kaur, A., &Singh,N.(2020). "Phenolic composition, antioxidant potential and health benefits of citrus peel."Food Research International, 132, 109114.

Soković. M, Marin, D. P., Brkić, D, Griensven, Leo J. L. D., &van Griensven. (2007). "Chemical Composite and Antibacterial Activity of Essential Oils of Ten Aromatic Plants against Human Pathogenic Bacteria."Global Science Book, Food 1(1), x-y.

Wang,C. Y., Chuang, C.Y., & Hsu, W.H.(2008). "The Flavonoids,Carotenoids and pectin content in peels of Citrus cultivated in Taiwan." Food Chemistry, 106,277-284.

Wong, C., Chug,B. L. H., Wing, C.K.., &Chen, F. (2006). "A Systematic Survey of antioxidant activity of 30 Chinesemedicinal plants using the ferric reducing antioxidant power assay,"Food Chemistry,97,705-711.

Yolanda, Y. P.,J. F., Enrique, A.,Danial, B. A., Botello., &Zamilpa, A. (2010). "Citrus limetta leaves extract antagonizes the hypertensive effect of angiotensin II."Journal of Ethnopharmacology,128, 611-614,2010.

Zheng,W., & Wang, S.Y. (2001). "Antioxidant Activity and Phenolic Compounds in Selected Herbs.J. Agric."Food Chem.,49, 5165–5170.

Zou,Z., Xi W., Hu Y.,Nie C., &Zhou,Z. (2016). "Antioxidant activity of citrus fruits."Food Chem.,2016, 196,885–896.

9

Biosorption: Heavy Metals Removal Through Biofilms and Microbial Polysaccharides

Anchal Singh, Mazia Ahmed, Pinki Saini* and Devinder Kaur

Centre of Food Technology University of Allahabad, Uttar Pradesh

Abstract

Water is very essential nutrient and total of only 3% pure water is present on earth. Some natural, anthropogenic and human activities discharge wastes, containing heavy metals into the different water resources and ultimately destroying the ecosystem. Numbers of methods are used for water purification like conventional (includes physical and chemical) and biological. Conventional methods are expensive, generate large quantities of sludge and inefficient at low concentrations of metal. Biological method includes biosorption and bioaccumulation, are eco-friendly and economic and have advantages over conventional methods. Biosorbent material includes microbial biomass, agro-wastes, and industrial by-products are used to treat heavy metal from liquid samples. Actually these biosorbent have number of functional groups on their surfaces which help in binding of metal from aqueous solution. In this article, biosorption of zinc, arsenic and copper are described through lactic acid bacteria that are not harmful, nonsporulating, fastidious, fermentative and generally regarded as safe. Many of these species are probiotic and producing exopolysaccharides and biofilms. It has great resistance to survive in gut environment and efficiency to biosorb metal from gut. It has some significant characteristics that represent a useful tool for decontamination of food and beverages from heavy metals. Biosorption mechanism and factors behind the process make these strains competent of removing these heavy metals.

Introduction

Comprising over 70% of the earth's surface, water is undeniably the most important natural resource of an ecosystem. Although, we also can't ignore the fact, pollution of water resources is common occurrence. Particularly, potable water has become highly affected. Water pollution or contaminants sources are two types direct sources and indirect contaminant/sources. Different industrial effluents, waste treatment plants and refinery companies are direct source of contaminants while acid rains, leaching of heavy metals from soil into groundwater are indirect source of contaminants. Heavy metals have higher density and higher atomic weight. Heavy metals include bismuth, tellurium, thallium, tin, antimony, arsenic, gold, silver, gallium, nickel, cadmium, chromium, cobalt, copper, lead, mercury, manganese, platinum, cerium, uranium, vanadium, iron, and zinc. These heavy metals are extended in the whole environment through several natural (soil erosion, spring waters, and volcanic eruptions) as well as anthropogenic activities (industrial processes, fossil fuel combustion and agricultural pesticides and herbicides). Each living beings are exposed to heavy metals through food chain (ingestion) or (inhalation). People living or working near industrial site are more prone towards the risk of exposure. Lifestyles and food habits can also impose higher risks of metal contamination like cigarette smoking is source of cadmium poisoning and shows different health implications. Among all heavy metals some are more toxic, included cadmium, chromium, lead, arsenic, mercury, cobalt and some are less or long-term effect of higher concentration of zinc and copper became toxic. These toxic metals bioaccumulate and lead to various health hazardous diseases and disorders even at very trace amount. So heavy metal contamination is a serious and emerging problem of whole ecosystem.

Today, maximum research interest has been focused on biological and economic way for removal of heavy metal pollution. So biosorption has emerged as an attractive waste management or metal remediation technique over other conventional methods. Biosorption is a quick, independent and metabolically passive process responsible for the selective sequestration of heavy metals by dead/inactive biomaterials or biomass. It is a simple physicochemical phenomenon resembling to conventional adsorption or ion exchange method.

The only difference is the nature of sorbent which in this case is the material of biological origin known as biosorbent. Biosorbents may be viewed as natural ion-exchange materials that primarily contain weakly acidic and basic groups, the chelation process being unspecific. These are the renewable biological materials responsible for the heavy metal removal from environment due to their metal sequestering properties (Wang and Chen 2009). Selection of biosorbent depends on the origin, availability and cost-effectiveness of biomass. They can be collected directly from the environment or specially developed by several modification processes (Dhankhar and Hooda, 2011; Vijayaraghavan and Yun, 2008).

Lactobacillus is a genus of gram positive, facultative anaerobic or microaerophilic, rod-shaped, non-spore forming bacteria. They are a major part of the lactic acid bacteria (LAB) group (i.e., they convert sugars to lactic acid). LAB has been widely used around the world to enhance quality like preservation, texture, flavour aroma and nutritional value of a large variety of products like milk, meat, cereals, vegetables and fruits. The lactic acid fermentation of milk is by far the best studied and known. A great variety of dairy products with different flavour, texture and health-promoting properties can be obtained from milk using different technologies and starter cultures. Important features of the starter are rapid acidification, microbial preservation of the milk, formation of specific flavours, texturing capacities and health benefits. LAB are able to produce numerous polysaccharides and the nature of it's dependent upon sugar units present in its surface. The polysaccharide is also known as exopolysaccharides (EPSs). Bacteria release two types of exopolysaccharides in form of slime and capsule. An adherent cohesive layer of polysaccharides are called capsular polysaccharides and the polysaccharides loosely attached or completely excreted into the environment are called slime (Cerning, 1994). EPSs also have a role in cell recognition, in adhesion to surfaces and in formation of biofilms facilitating the colonisation of various ecosystems (Looijesteijn et al., 2001; Hugenholtz, 2001; Whitfield and Valvano, 1993).

Biofilm may be defined as a community of microorganism grown in surface release their genetic material and exopolysaccharides (Costerton,1995). Biofilm has high biosorption of heavy metals. It may be single microbial species or multiple microbial species and able to form an array of biotic and abiotic EPS surfaces. According to National Institutes of Health (NIH) about 65% of all microbial infections, and 80% of all chronic infections are associated with biofilms. Biofilm formation is a multi-step process starting with attachment to a surface then formation of micro-colony that direct the arrangement of 3D structure and finally ending with maturation followed by detachment (Jamal, 2015). Extracellular polymeric substances, derived from metabolism of microbial cells in the culture medium. Numerous studies have now shown that carbohydrates and proteins are found to be the major components of EPS and various metallic elements (e.g. Fe, Mg, K, Mn) are present in the EPS as well. EPS was considered as an organic product with mineral cations bound (d'Abzac et al., 2009).

Approximately 30 species of lactobacilli are described as EPS producers. Among them, the best known are *L. casei, L. acidophilus, L. brevis, L. curvatus, L. delbrueckii bulgaricus, L. helveticus, L. rhamnosus, L. plantarum, L. johnsonii,* etc. They are principally cultivated between 30 and 37 °C on rich media as MRS (Man Rogosa Sharp), milk or milk derivatives. The culture is generally conducted without agitation due to their aero-anaerobic status.

Heavy metals

Heavy metals are occurring naturally and through anthropogenic activity in our environment and are present in whole ecological unit (Krishna et al., 2016). Natural source involves volcanic eruptions, landslides, spring waters, soil erosion, and biomagnifications. Anthropogenic activities include waste treatment plants, fossil fuel combustion, smelting, mining, agricultural activities (Florea et al., 2004). Metals having density more than $5gm/cm^3$ with higher atomic weight greater than 40.04 (Duffus, 2002). Heavy metals such as cadmium, chromium arsenic, tin, tellurium, thallium, gold, lead, cobalt, mercury, bismuth, manganese, nickel, molybdenum, uranium, platinum, gold, silver, vanadium, iron, copper, and zinc etc. Some of them are essential micro minerals like cobalt, copper, zinc, iron, selenium, nickel. Very less quantity of these metals are vital and performing different physiological and biochemical functions in body but excess amount lead chronic or acute toxicity. In human body heavy metals are transported through blood stream into the cell and tissues. The transportation of metal from different body parts has been done by metals binding lipid protein and nucleic acid due to their complex structure and availability of free functional groups. Heavy metal form complex with biomolecules and denature the functioning of cell. Malfunctioning of cell affect activity of different organs like it may develop carcinogenicity, damage brain, liver, lung, heart and kidney (Monisha et al., 2014). Chronic or acute exposures of heavy metals lead to develop neurotoxicity and degenerative diseases.

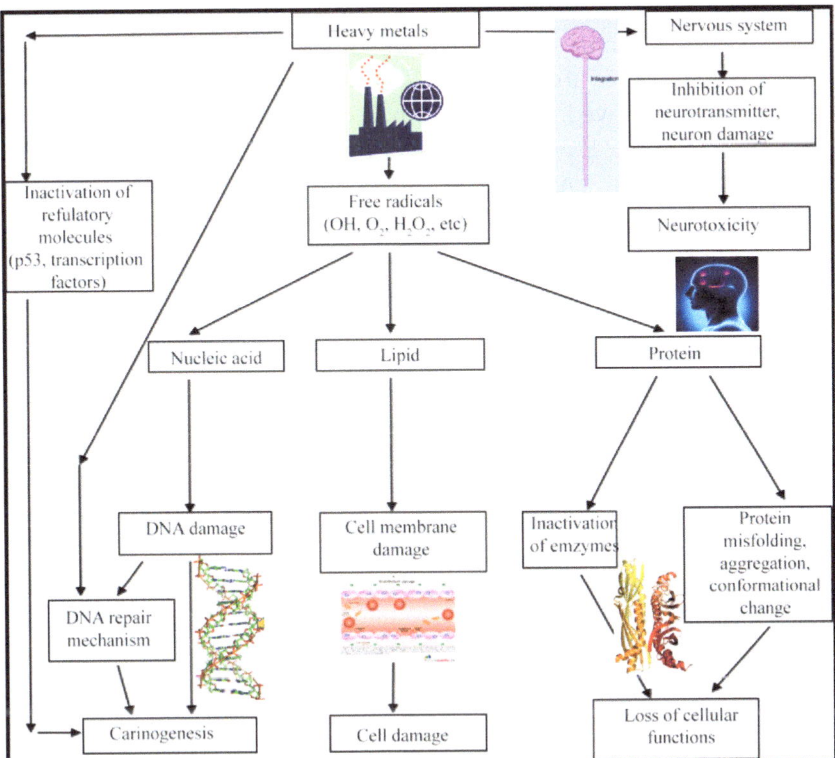

Fig. 1: Heavy metals and their carcinogenicity
(Source: Intech book; *Mechanism and Health Effects of Heavy Metal Toxicity in Humans)*

Heavy metals, toxicity and their impacts on human health

The heavy metals are toxic substance found everywhere in the ecological unit. The ecosystem exposed through the heavy metals by various natural and anthropogenic activities. Heavy metals present in earth crust in form of ores. Mining activity provides separate minerals in form of sulphide or oxide like, cadmium, arsenic, iron, chromium, lead, zinc, copper, gold, nickel, silver and cobalt. In this article, will discuss the various sources, occurrences, and toxicity of zinc, arsenic, copper and their removal through lactic acid bacteria and its exopolysaccharides by biological process biosorption. The descriptions of these metals are as follows:

Zinc

Zinc is an essential micronutrient for human metabolism that catalyses more than 100 enzymes, facilitates protein folding, and helps regulate gene expression. Zinc is the second most abundantly distributed trace element in the body after iron (King, 2006). Zinc chemically may be defined through the symbol Zn with atomic number is 30 and atomic mass 65.37. Zinc is a slightly brittle metal at 25 to 30 degree Celsius with blue-silvery appearance when oxidation is removed. This is belonging to group 12 elements of the periodic table. In some aspects, zinc is chemically similar to magnesium, like both elements are in same size and have same oxidation state (+2). Zinc has five stable isotopes. Zinc comprise about 0.0075% of total earth's crust and 24th most abundant element. The element is usually found in bound form linked with metals like copper and lead. Zinc added with sulphur and other heavy metal so called chalcophile. Electronic configuration of zinc is $[Ar]3d^{10}4s^2$ and is a member of the group 12 of the periodic table. It is a moderately reactive metal and strong reducing agent. The surface of the pure metal tarnishes quickly, eventually forming a protective passivating layer of the basic zinc carbonate. Zinc chemistry is similar to the chemistry of the late first-row transition metals, nickel and copper, though it has a filled d-shell and compounds are diamagnetic and mostly colourless.

Zinc is an essential trace element that can cause symptoms of deficiency and can be toxic when exposures exceed physiological needs. Zinc toxicity is a medical condition in which exposure of zinc is more than the normal i.e., above 50 mg that is considered as toxic for human. As per the guidelines of ICMR, the recommended allowance of zinc in man is 12 mg/day and 10mg/day in women. Excessive absorption of zinc inhibits the absorption copper and iron. The free form of zinc ion is used as bactericidal activity. People are exposed to zinc primarily from food, although oral exposure can become excessive through non-dietary sources. Certain occupational exposures can be hazardous. There are three major routes of entry for zinc into the human body; by inhalation, through the skin, or by ingestion (toxicological profile for zinc). Metal fume fever is mostly known toxicity developed by inhalation of zinc oxide smoke. MFF is an acute syndrome or industrial disease which is commonly due to inhalation of more than one millimicron particle released from zinc smelting or welding (Vogelmeier 1987; Plum et al., 2010). Sign and symptoms of MFF are fatigue, fever, upper and lower respiratory distress, nausea, and muscular dystrophy (Rohrs, 1957). This fever is not critical and the respiratory illness disappears within 1 to 4 days.

Ingestion of zinc directly affects the gastrointestinal tract, before distributed throughout the whole body. Oral ingestion of zinc develops numerous GIT symptoms. Brown et al. (1964) reported that use of galvanized containers in storage of food and beverages increase zinc content of product. Leaching of zinc from galvanized coating is depending upon acidic nature of food products. The resulting Sign and symptoms included abdominal cramps, nausea, vomiting, and diarrhoea, brown.

Arsenic

Arsenic defined as natural occurring element of the earth's crust and is widely distributed throughout the environment in the air, water and land. It is highly toxic in its inorganic form. People are exposed to higher concentration of inorganic arsenic through drinking contaminated water, using contaminated water in food preparation and irrigation of food crops, industrial processes, eating contaminated food and smoking tobacco. Long-term exposure to inorganic arsenic, mainly through drinking-water and food, can lead to chronic arsenic poisoning. Skin lesions and skin cancer are the most characteristic effects. Arsenic is a natural occurring element that an individual typically encounters every day in food, water, soil, and air. Arsenic is generally copious element on earth. Its inorganic forms like arsenite and arsenate are fatal to human and other living organisms. Arsenic exposure may take place through different reasons, consist of smelting, paints, wood preservatives and microelectronic industries, pesticides, herbicides, and fungicides (Sauvé's, 2014).

Arsenic is widely dispersed in earth's crust (Garelick et al., 2009). Arsenic occurs in many minerals, usually in combination with sulphur and metals, but also as a pure elemental crystal. Arsenic is a metalloid. It has various allotropes, but only the grey form, which has a metallic appearance, is important to industry. Arsenic found in water is almost entirely in the inorganic form and can be stable as both arsenite and arsenate, trivalent and pentavalent inorganic arsenicals, (Saxe et al., (2006). Arsenic, a metalloid and naturally occurring element, is one of the most abundant elements in the earth's crust and is found throughout our environment. Arsenic can attach to very small particles in the air, stay in the air for many days, and travel long distances.

Arsenic is a heavy metal with the atomic number 33 atomic mass 74.9216 and chemical symbol As. Arsenic found in two oxidation states: a trivalent form is called arsenite (As_2O_3) and arsenate (As_2O_5) a pentavalent form. Toxicity of arsenite is 60 times more than arsenate. Organic arsenic is non-toxic while inorganic arsenic is toxic. Exposure of arsenic may develop acute or chronic toxicity depending upon rout and exposure. Acute arsenic poisoning can damage the blood vessels, tissue lining of gastrointestinal tract, heart and brain. Chronic arsenic toxicity is termed as arsenicosis lead pigmentation and keratosis of skin (Martin, 2009). Short term exposure of arsenic can lead nausea, vomiting, anaemia, leukemia, vascularisation of blood vessels, abnormal heart beat which develop pricking feeling in hands and feet (Ratnaike, 2003). Long-term exposure can cause to the formation of skin lesions, pulmonary disease, CVD, neurological problems, diabetes mellitus, and hypertension (Huy et al.,

2014). Chronic effect of arsenic cause permanent changes and vascularisation of critical organs that may lead to death. It develop different cancers like skin cancer, bladder cancers, cancer of lung, liver cancer, and kidney cancers (Tuyet et al., 2014). Carcinogenicity developed by arsenic is DNA damage and generation of reactive oxygen species (Martinez et al., 2011; Aaseth et al., 2017).

Copper

Copper (Cu) is an integral part of many important enzymes involved in a number of vital biological processes. Copper is found in a different cells and tissues but the highest concentrations present in the liver and brain (Turnlund, 1998). In biological systems copper is exist in cupric form (Cu^{++}), and generally found, bound with protein and enzymes. (Divertie et al., 1982). The copper containing enzyme is essential for cross-linking of connective tissue collagen and elastin. The copper containing protein, ceruloplasmin is responsible for transportation of copper from the interstitial lumen to sites of erythropoiesis. Copper is essential for the formation and maintenance of myelin sheath and the formation of melanin pigment in skin, hair, and eyes. Copper is a trace mineral and chiefly present in brain, liver, and kidney. Copper is produced in massive stars (Romano et al., 2007) and is present in the Earth's crust in a proportion of about 50 parts per million (Emsley and John, 2003). In nature, copper occurs in a variety of minerals, including native copper, copper sulfides such as chalcopyrite, bornite, digenite, covellite, and chalcocite. Copper is the 26th element in abundance in the crust of the earth and is the 29th element in the periodic table with 2 stable and 9 radioactive isotopes.

Copper is a transition metal having atomic mass of 63.54 Daltons. Due to high electrical conductivity, low corrosion, alloy forming ability, high thermal conductivity it has wide range of industrial application (Lisa et al., 2003). Copper is an essential constituent of white gold and other alloys used for making artificial jewellery, dental products, and in many other cosmetics (Okereke et al., 1972). Copper has high nuclear charge with high ionization potential (Georgopoulos et al., 2006). Copper toxicity is very infrequent in humans, the homeostatic of human can control of copper reactivity within cell. Presence of free copper is extremely low in cells or in the different body parts. copper always exists bound form mainly with proteins. Excess intake of copper and disorder in copper metabolism are responsible for developing Wilson's disease. Recommended intake of copper not more than 10mg/day is established (Trumbo et al., 2001). Toxicity of Copper is generally higher in neonates and infants due to an immature biliary excretion which enhanced the copper absorption in intestinal. Result of copper overloading is responsible in impairment of biliary copper excretion which may leads biliary cirrhosis and biliary atresia.

Copper toxicity mainly seen in the liver, because after enter into the blood stream it is firstly deposited in liver. Copper toxicity may lead primarily liver cirrhosis with episodes of haemolysis further it damages renal tubules, the brain, and other organs severe condition lead to hepatic necrosis, coma and death (Winge and Mehra, 1990). Chronic copper toxicity is well documented in dialysis patients that receive dialysis via copper tubing (Klein et al., 1972), long term contact with copper containing pesticides (Winge and Mehra, 1990) and in infants with chronic enteral or total parenteral nutrition (Beshgetoor and Hambidge, 1998). At early stages this poisoning may result in weakness, lethargy, and anorexia (Semple et al., 1960; Winge and Mehra, 1990) Research reported that ingestion of copper contaminated foods and water may develop of acute GIT symptoms like lesion in lining of epithelium of intestine, hepatocellular necrosis in the liver, and acute tubular necrosis in the kidney (Barceloux, 1999; Knobeloch et al., 1998; Spitalny et al., 1984).

Lactic acid bacteria

Lactobacillus is a genus of Gram positive, facultative anaerobic or microaerophilic, rod-shaped, non-spore forming bacteria. They are a major part of the lactic acid bacteria group (i.e., they convert sugars to lactic acid). Lactic acid bacteria are a group of related bacteria that are able to produce lactic acid after fermentation of carbohydrate (Vijayarahavan and Yun, 2008; Masood et al., 2010). LAB is broadly used in the production of fermented food product, such as yogurt (*Streptococcus* spp. *and Lactobacillus* spp*.),* cheeses (*Lactococcus* spp.), and sauerkraut (*Leuconostoc* spp.). They are food grade organisms possessing the generally regarded as safe (GRAS) status. LAB is used in food industry because their growth lowers both the CHO content of the foods they ferment and the pH due to lactic acid production. The pH may drop to as low as 4.0, low enough to inhibit the growth of most other micro-organisms including the most common pathogen, thus allowing these foods prolonged shelf-life. (Zehra et al., 2008).

Lactic acid bacteria includes a wide group of genus that includes *Lactobacillus, Streptococcus, Enterococcus, Lactococcus, Bifidobacterium, and Leuconostoc* (Masood et al., 2010). They are gram positive, non-sporulating, nonmotile, facultative anaerobes that gained energy through sugar fermentation and produce lactic acid as end products. They required different nutrients like purines, pyrimidines, vitamins, and amino acids for their growth and development and take part in biosynthesis of vitamin B-12 and folic acids

(Srivastava et al., 2020). The order LAB have largest genus *Lactobacillus* which includes approximately 80 species which are widely present and used in production of cheese, yogurt, beer, juices, wine, pickle, sauerkraut, and sausage (Prescott et al., 2002; Azizpur, 2009; Masood et al., 2010). Different researchers have evaluated the safety limit of lactic acid bacteria in human beings. This genus also detected some side effects like Lactobacillemia and antibiotic resistance due to excessive colonization and some time it produce biogenic amines during fermentation or by fermented products cause unwanted adverse effects. These effects are rarely so it is considered as safe (Bernardeau et al., 2008).

Most LAB are used as probiotics. The term probiotics may be defined as "Living organisms" which provide numerous health benefits (Vijayarahavan and Yun, 2008). LAB is useful for human being and animals in many aspects. LAB are useful probiotics in different health conditions like prevention of diarrhoea, in case of lactose intolerance, in treatment of ulcer, stimulate immunity of both intestinal and systemic level, reduction of colon cancer, antifungal activity.

Microbial balance plays a significant role for maintaining the intestinal homeostasis. Intake of live LAB through dairy products have many beneficial effects on GIT of humans. These live organisms are used to correct lactose malabsorption, treat viral and drug induced diarrhoea, irritable bowel and inflammatory bowel syndrome, maintain blood glucose level and total maintenance of gut health. Lactic acid bacteria employed the beneficial role in renovation of intestinal microflora, strengthening the intestinal barrier, eradication of intestinal pathogens, promote the process of phagocytosis, and formation of anti-inflammatory products (Hove et al., 1999; Harish and Varghese, 2006; Heyman, 2000).

Exopolysaccharide

Microorganisms produce different polysaccharides that are classified according with their location and stress related to the cell habitat. The polysaccharides that are excreted outside the cell wall are called extracellular polysaccharides or exopolysaccharide. The exopolysaccharides released in the environment in two types capsular and slime. Capsule is intact form but slime is loosely excreted into the environment (Cerning, 1994). During growth phase, produced capsular exopolysaccharides are permanently attached to the microbial cell surface while slimes are secreted from the microbial cell into their surroundings (Laws et al., 2001). The bacterial EPSs are not used as energy sources by the producer microorganism. They probably have a protective function in the natural environment, e.g. against desiccation, phagocytosis and predation by protozoa, phage attack, antibiotics or toxic compounds and osmotic stress.

EPSs also have a role in cell recognition, in adhesion to surfaces and in formation of biofilms facilitating the colonisation of various ecosystems (Vries et al., 2001; Whitfield and Valvano, 1993).

EPSs are long-chain branched polysaccharides consist of repeating units of oligosaccharides and sugars. The sugar and oligosaccharides generally present are glucose, galactose, inositol and rhamnose, in different relative amounts (Vuyst et al., 1999). Microbial exopolysaccharides are classified into two groups– homopolysaccharides like cellulose, mutan, dextran, alternan, curdlan, levan and pullulan, and heteropolysaccharides like xanthan and gellan. EPS are polymeric substances excreted by the microorganisms when growing in environment and play a major role in the formation of bio-films to grow on the substrates in environments where normally others cannot colonise (Sutherland, 2001). In some species, EPS play an essential role in imparting antibiotic resistance to the organism by denying permeability to the antibiotics (Costerton et al., 1987).

The EPS produced by LAB improve the texture and viscosity of naturally fermented milk products and prevents synergies. When EPS is added to food it exhibits function like thickeners, stabilizers, emulsifier, gelling agents and water binding agents. Bacterial EPS show enormous diversity. In addition to their natural protective role, EPS produced by LAB is found to shield the cell against technological stresses during manufacturing of fermented milk products. Application of these polymers can be done for human usage, medical, cosmetics, pharmaceutical, dairy product and other forms of industrial and environmental aspects. EPS production enhances the probiotic potential of bacteria. Thus, EPS plays a significant role in the colonization of LAB to various ecosystems by facilitating the colonization of bacteria into intestinal mucosa that enhance the immunity of host.

Exopolysaccharides (EPSs) secreted from lactic acid bacteria (LAB) have wide application in different food sectors like it improve rheological property, texture and 'mouthfeel' of fermented milk and milk products. EPSs are the better alternatives to fulfil the consumer demand for developing a product with low fat, sugar and additives, as well as cost effective for development of rich smooth and creamy yoghurt product (Jolly et al., 2002). Even though there is no own taste of EPS produced from LAB but they produce taste and enhance the sensitivity of taste when they present in milk and milk products (Duboc and Mollet, 2001). It has proved that EPS have beneficial effect in gastrointestinal tract through enhancing colonization of probiotics by providing a substrate substance and also work as prebiotics (German et al., 1999). Additionally, the exopolysaccharides of LAB have antitumor effects, cholesterol lowering and immunomodulating activity (Hosono et al., 1997; Nakajim et al., 1992; Chabot et al., 2001).

Biofilm

Biofilms can be defined simply and broadly as communities of microorganism that are attached to the surface. A concerted effort to study microbial biofilms began only two decades ago with the rediscovery that, in natural aquatic systems, bacteria are found predominately attached to surfaces. (Geesey et al.,1977). Biofilm has low cost production and good biosorption capacities. It is better option in the field of water treatment as compare to other conventional methods (Chubar et al., 2008; Gadd, 2009). However, the research related to utilization of naturally formed biofilm in environment for biosorption of toxicant is rarely accomplished. Many studies focused on the adsorption of heavy metal by laboratory grown bacterial biofilm of same strain or mixed culture (Almaguer-Cantú, 2011; Fang et al., 2010; Hawari et al., 2005, Borrock et al., 2004; Johnson et al., 2006 and Joo et al., 2010).

Biofilm polymer are one of the major parts of biofilm (Sutherland, 2001). The polymer is rich in functional groups that provide binding sites and improve the biding from aqueous solutions (Hiraki et al., 2009 and Tsuchiya et al., 2009). Biofilm development act as an adsorbent for metal ions in the existent surroundings, this is established by correlation between adsorption behaviour of biofilm and polymer property. toxicant like heavy metal ions is one of the key contaminants of water pollution (Vijayaraghavan et al., 2008 and Volesky, 2007). Lithium ion is one of new contaminants added in list of heavy metals of more concerns. Due to use in different industrial activity lithium is consumed as a key compound in batteries, high-performance grease, flux for welding, heat-resistant ceramics and pharmaceuticals (Kaneko et al., 1990; Miyai et al., 1978). Lithium used in manufacture of nuclear fusion furnace (Dang et al., 1978, and Harlty et al., 1978). Thus, the recovery of lithium is an important subject (Tsuruta, 2005).

Biosorption

There are various techniques have been used for the treatment of dyes/metal bearing waste or effluents. Today's most of the research has payed attention on biological methods for waste management (Vijayaraghavan, 2003). The main advantages to use biotechnologies for the removal of pollutants is because of it can be performed at the contaminated site, eco-friendly means no secondary pollution and also these methods are cost effective.

Biosorption and bioaccumulation have been employed to efficient technique as compare to conventional methods for the removal of heavy metals (Volesky and Holan, 1995). There is some difference between the terms "bioaccumulation" and "biosorption" depending upon the state of the biomass. Bioaccumulation is the phenomenon of living cells while the term biosorption are mechanism of dead biomass. Bioaccumulation may be defined as the uptake of pollutants through metabolically mediated active absorption process across the cell membrane of living cells (Malik, 2004). Biosorption can be defined as the metabolism-independent processes passive absorption of toxic substances through dead or inactive biological materials or biomass through it's cell wall. Removal of toxicants through biological methods is called biosorbent include plant derived materials, agriculture or industrial wastes, microorganisms biofilms, exopolysaccharides, biopolymers, and so on.

Biosorption is a rapid reversible process implicated in binding of metal ions through functional groups present on the surface of the biosorbent materials. The process required aqueous or liquid phase containing metals or toxicants in solutions form and biosorbent or solid material for adsorption of metals through liquid solution (Davis et al., 2003). Thus, it is highly recommended process due to easy procedure, no requirement of additional nutrient, less generation of sludge, economic, high efficiency, regeneration of biosorbent materials, and no need of chemical oxygen demand (COD) of water, that are essential and limiting factor for majority of the conventional techniques (Chojnacka, 2010). Most potent technique biosorption can able to remove toxicants even in dilute concentrations and effective at removal of very less concentration of metals or toxicants even at ppb levels.

Mechanism

Biosorption is a rapid mechanism of metal uptake on the cell surface, known as extracellular binding. In this process, metal adheres to surface molecules such as S-layer protein (SLP) (Gerbino et al., 2015). The binding occurs by any one or a combination of the processes includes: physical adsorption, van der Waals forces, ion exchange, complexation or inorganic microprecipitation (Srinath et al., 2002).

The first step, physical adsorption is associated with the presence of Van der Waal's forces (Crowell, 1966). Cell wall is the first component that comes in contact with the metal ions and also acted as a defence against metal toxicity. Therefore, the chemical makeup of the microbial cell wall is an important factor responsible for the adsorption of metal ions on the biomass' surface, which varies among different groups of micro-organisms results in significant difference in the metal binding type. Bacteria make excellent biosorbents because of their high surface to volume ratios and presence of potentially active chemisorption sites (teichoic acids) in their cell walls (Beveridge, 1989; Vijayaraghavan and Yun, 2008). Gram-positive bacterial cell wall has teichoic acids containing phosphoryl and hydroxyl groups, which deprotonate into negative charge in alkaline condition acting as cation adsorption site. In gram-negative bacteria, the phosphate groups within lipopolysaccharides (LPSs) and phospholipids have been demonstrated to be the primary sites for metal interaction (Remacle, 1990).

Huang and Liu (2013) and Davis et al. (2003) reported that surface binding is the principle phenomenon for biosorption of metal ions, which occurs due to ion-exchange mechanism of biomass. Ion exchange involves competition between protons and metal cations for the binding sites. Matheickal et al. (1997) reported that dominant mechanism of Cu (II) biosorption by *Ecklonia radiata* is ion exchange mechanism involves exchange of $Ca^{2+}$and Mg^{2+}ions present in their cell wall. The metal removal from solution may also take place by complex formation on the cell surface after the interaction between metal cations and active groups present on the cell surface (Han et al. 2006). Gadd and Griffiths (1978) reported that complexation of metals such as copper, cadmium, and zinc is possible with polygalacturonic acid, an important constituent of the outer layers of bacterial cells. Complexation was found to be the only mechanism responsible for calcium, magnesium, cadmium, zinc, copper and mercury accumulation by *Pseudomonas syringae* (Cabral, 1992).

Advantages of biosorption

Following are given the advantages of biosorption over conventional metal removal methods (Aksu et al., 1992 and Abbas et al., 2014).

1. Cheaper production of biomass (bacteria or fungi).
2. Use of biomass for removal of heavy metals.
3. Multiple heavy metals uptake at a time
4. Treatment of large volumes of wastewater.
5. No need for chemical additions as highly selective for uptake and removal of specific metals.
6. Functional over wide range of conditions including temperature, pH, presence of other metal ions, etc.
7. Easy and cheaper desorption of metals attached to biomass.
8. Reduced volume of waste or toxic materials production.

Conclusion

Biosorption is an eco-friendly and cheap method of removing metals from the environment. From studies it is inferred that the microbial cells retain the activity in an optimum pH and also depend on other factors. This help in identifying the suitable biosorbent for biosorption process. There are some drawback of biosorption i.e. it is not applicable at broad level, developing a commercial biosorbent similar to an ion resin is essential, which only need low growth conditions and is easy to carry out large-scale industrial cultivation and help to reduce the cost of biosorbent. Totally, the successful application of biosorption requires the joint efforts of workers from different backgrounds, so that biosorbents such as lactic acid bacteria produced exopolysaccharide and biofilms can play an important role in the prevention and control of heavy metal pollution.

References

Almaguer-Cantú, V., Morales-Ramos, L. H., & Balderas-Rentería, I. (2011). Biosorption of lead (II) and cadmium (II) using Escherichia coli genetically engineered with mice metallothionein I. Water Science and Technology, 63(8), 1607-1613.

Borrok, D., Fein, J. B., & Kulpa, C. F. (2004). Proton and Cd adsorption onto natural bacterial consortia: tcsting univcrsal adsorption bchavior. Gcochimica ct Cosmochimica Acta, 68(15), 3231-3238.

Cerning, J. (1994). Polysaccharides exocellulaires produits par les bactéries lactiques.

Chubar, N., T. Behrends and P.V. Cappellen. (2008). Biosorption of Metals (Cu^2, Zn^2) and Anions (F, H PO) by Viable and Autoclaved Cells of the Gram-negative Bacterium Shewanella putrefaciens. Colloids and Surfaces B: Biointerfaces; 65:126 133.

Dang, V. D., and Steinberg, M. (1978). Preliminary design and analysis of recovery of lithium from brine with the use of a selective extractant. In Lithium Needs and Resources (pp. 325-336). Pergamon.

Dhankhar, R., and Hooda, A. (2011). Fungal biosorption–an alternative to meet the challenges of heavy metal pollution in aqueous solutions. Environmental Technology, 32(5), 467-491.

Fang, L., Cai, P., Li, P., Wu, H., Liang, W., Rong, X., and Huang, Q. (2010). Microcalorimetric and potentiometric titration studies on the adsorption of copper by P. putida and B. thuringiensis and their composites with minerals. Journal of Hazardous Materials, 181(1-3), 1031-1038.

Gadd, G. M. (2009). Biosorption: critical review of scientific rationale, environmental importance and significance for pollution treatment. Journal of Chemical Technology & Biotechnology: International Research in Process, Environmental & Clean Technology, 84(1), 13-28.

Gaetke, L. M., and Chow, C. K. (2003). Copper toxicity, oxidative stress, and antioxidant nutrients. Toxicology, 189(1-2), 147-163.

Hartley, J. N., Gore, B. F., and Young, J. R. (1978). Potential lithium requirements for fusion power plants. In Lithium Needs and Resources (pp. 337-346). Pergamon.

Hawari, A.H., and C.N. Mulligan. Biosorption of lead (II), cadmium (II), copper (II) and nickel (II) by anaerobic granular biomass. Bioresour. Technol. 2005; 97:692 700.

Hiraki, A., Tsuchiya, Y., Fukuda, Y., Yamamoto, T., Kurniawan, A., and Morisaki, H. (2009). Analysis of how a biofilm forms on the surface of the aquatic macrophyte Phragmites australis. Microbes and environments, 0908100108-0908100108.

Johnson, K. J., Szymanowski, J. E., Borrok, D., Huynh, T. Q., and Fein, J. B. (2007). Proton and metal adsorption onto bacterial consortia: Stability constants for metal–bacterial surface complexes. Chemical Geology, 239(1-2), 13-26.

Joo, J. H., Hassan, S. H., and Oh, S. E. (2010). Comparative study of biosorption of Zn2+ by Pseudomonas aeruginosa and Bacillus cereus. International Biodeterioration & Biodegradation, 64(8), 734-741.

Kaneko, S., and Takahashi, W. (1990). Adsorption of lithium in sea water on alumina—magnesia mixed-oxide gels. Colloids and surfaces, 47, 69-79.

Kurniawan, A., and Yamamoto, T. (2013). Biofilm polymer for biosorption of pollutant ions. Procedia Environmental Sciences, 17, 179-187.

Looijesteijn, P. J., Trapet, L., de Vries, E., Abee, T., and Hugenholtz, J. (2001). Physiological function of exopolysaccharides produced by Lactococcus lactis. International Journal of Food Microbiology, 64(1-2), 71-80.

Malik, A. (2004). Metal bioremediation through growing cells. Environment International, 30(2), 261-278.

Masood, M. I., Qadir, M. I., Shirazi, J. H., and Khan, I. U. (2011). Beneficial effects of lactic acid bacteria on human beings. Critical Reviews in Microbiology, 37(1), 91-98.

Miyai, Y., Kanoh, H., Feng, Q., and Ooi, K. (1995). Intraparticle diffusion coefficient of lithium on granulated adsorbent of manganese oxide in seawater. Nippon Kaisui Gakkai-Shi, 49(6), 347-351.

Muhsin, J., Ufaq, T., Tahir, H., and Saadia, A. (2015). Bacterial biofilm: its composition, formation and role in human infections. J Microbiol Biotechnol, 4, 1-14.

Ozdemir, S., Kilinc, E., Poli, A., Nicolaus, B., and Guven, K. (2009). Biosorption of Cd, Cu, Ni, Mn and Zn from aqueous solutions by thermophilic bacteria, Geobacillus toebii sub. Sp decanicus and Geobacillus thermoleovorans sub. spstromboliensis: Equilibrium, kinetic and thermodynamic studies. Chemical Engineering Journal, 152(1), 195-206.

Plum, L. M., Rink, L., and Haase, H. (2010). The essential toxin: impact of zinc on human health. International journal of environmental research and public health, 7(4), 1342-1365.

Pokrovsky, O. S., Feurtet-Mazel, A., Martinez, R. E., Morin, S., Baudrimont, M., Duong, T., and Coste, M. (2010). Experimental study of cadmium interaction with periphytic biofilms. Applied Geochemistry, 25(3), 418-427.

Quintelas, C., Rocha, Z., Silva, B., Fonseca, B., Figueiredo, H., and Tavares, T. (2009). Removal of Cd (II), Cr (VI), Fe (III) and Ni (II) from aqueous solutions by an E. coli biofilm supported on kaolin. Chemical Engineering Journal, 149(1-3), 319-324.

Silva, R. M. P., Rodríguez, A. Á., De Oca, J. M. G. M., and Moreno, D. C. (2009). Biosorption of chromium, copper, manganese and zinc by Pseudomonas aeruginosa AT18 isolated from a site contaminated with petroleum. Bioresource Technology, 100(4), 1533-1538.

Srivastava, U., Saini, P., and Singh, A. (2020). Effect of natural fermentation on antioxidant activity of pearl millet (Pennisetum glaucum). Current Nutrition & Food Science, 16(3), 306-313.

Sutherland, I. W. (2001). The biofilm matrix–an immobilized but dynamic microbial environment. Trends in Microbiology, 9(5), 222-227.

Tsuchiya, Y., Ikenaga, M., Kurniawan, A., Hiraki, A., Arakawa, T., Kusakabe, R., and Morisaki, H. (2008). Nutrient-rich microhabitats within biofilms are synchronized with the external environment. Microbes and Environments, 0901060040-0901060040.

Tsuruta, T. (2005). Removal and recovery of lithium using various microorganisms. Journal of Bioscience and Bioengineering, 100(5), 562-566.

Vijayaraghavan, K., and Yun, Y. S. (2008). Bacterial biosorbents and biosorption. Biotechnology Advances, 26(3), 266-291.

Volesky, B. (2007). Biosorption and me. Water Research, 41(18), 4017-4029.

Wang, J., and Chen, C. (2009). Biosorbents for heavy metals removal and their future. Biotechnology Advances, 27(2), 195-226.

Whitfield, C., and Valvano, M. A. (1993). Biosynthesis and expression of cell-surface polysaccharides in gram-negative bacteria. In Advances in Microbial Physiology, Vol. 35, pp. 135-246. Academic Press.

Yamamoto, M., Murai, H., Takeda, A., Okunishi, S., and Morisaki, H. (2005). Bacterial flora of the biofilm formed on the submerged surface of the reed Phragmites australis. Microbes and Environments, 20(1), 14-24.

10

Nutraceutical Potential of Food Byproduct and Its Utilities

Pragya Mishra[1], Raghvendra Raman Mishra[2], Neelam Yadav[3]

[1]Food Processing and Management, DDU Kaushal Kendra, RGSC Banaras Hindu University, Uttar Pradesh

[2]Medical Laboratory Technology, DDU Kaushal Kendra, RGSC, Banaras Hindu University, Uttar Pradesh

[3]Centre of Food Technology, University of Allahabad, Prayagraj Uttar Pradesh

Abstract

The waste utilization as derivative from different food processing sector has been emerged as the main challengeable problem around the world due to the generation of by-products at large quantities consisting seeds, peels, rags, unused flesh, bran, husk, germ, stalk etc. during various processing chain steps. Though these food by-products are potentially rich source of natural compounds which finds its wide utilization in different industries as affordable, cost-effective, novel ingredient comprising dietary fiber, phytochemicals, antioxidants, pectin, enzymes, organic acids, food additives, essential oils etc. subsequently after extractions, purified fractionation and fermentations methods. These food by-products also have pharmaceutical and nutraceutical activity. A nutraceutical is like a medicine or food grade product, which not only provides physical benefits and nutritional supplements but also protects against various chronic diseases. This food processing waste utilization should be a topic of interest, by means of two ways, i.e., through waste quantity reduction and increased value-added food products production and also as renewable energy source. The purpose of this study is to highlight the opportunities to utilize food by-products from various food processing industries and to encourage the vital exploitation of the by-products containing large number of bio-active compounds having nutraceutical potential.

Introduction

Food processing wastes are the finish products attained from numerous food processing industries that have neither been recycled it nor used for any other purposes. Amongst all food wastes the agri-horticultural wastes contributes a big share in municipal landfills owing to their high degree of biodegradability. The statistical archives of United Nations Environment Programme (UNEP) demonstrate that every year one third of the food produced for human consumption throughout the world, of which, almost 1.3 billion tons of food is vanished or unexploited. The Food and Agriculture Organization (FAO) of the United Nations estimates that, one-third of all food produced is either lost or wasted every year. The food waste produced amounts to 1.3 million tons a year (Priedniece et al. 2017). The waste comprises 45 percent of fruit and vegetables, 35 percent of fishes and seafoods, 30 percent of cereals, 20 percent of dairy-based products and 20 percent of meat and its product (Lipinski et al. 2013). Waste products (approximately 192 million tons annually) from fruit and vegetable processing are difficult to manage. The waste gained from fruits and food processing industries are tremendously varied owed to the usage of inclusive fruits and vegetables variety, the wide array of procedures and the diversified product. For example, tropical and subtropical fruits processing contribute major proportions of by-products as compared to the temperate fruits. Due to growing manufacture and processing of fruits and vegetables-based products, disposal represents a rising problem subsequently the plant material is more prone to microbial attack and spoilage, thus limiting its further exploitation. Conversely, drying costs, storage, distribution and shipment of by-products are economically restrictive aspects. Food processing wastes are the finish products of immeasurable food processing industries that have no usages or not utilized properly for other purposes. The agro-industrial waste is most often utilized as feed, fodder or as fertilizer. The plant foods processing consequences in the manufacture of massive amount of by-products containing large quantities of bioactive compounds such as polyphenols, phytic compounds, antioxidants antimicrobial compounds which may potentially used for enhancing the foods storage stability by preventing lipid oxidation. The use of post-production waste for energetic purposes allows to the minimization of the waste quantity with reduced total cost of successive business. Plentiful researches have been conducted for the extraction of bioactive ingredients like phytochemicals, antioxidants, dietary-fiber, food ingredients like pectin, natural colour, vitamins, minerals, antibiotics and enzymes to one side and on the other hand fruits, vegetables, cereals, pulses milling, tea, coffee, cocoa and dairy processing units for production of ethanol or biogas generation (Schieber et al. 2001)

Nutraceuticals

Food and drugs obtained from natural resources have attained a fairly substantial part in human and public healthcare system all over the world. Human curiosity and exploration for specific constituents of plants, animals' minerals and microbial origin which are advantageous to inclusive health have cost creating of terminologies such as functional food or nutraceuticals. Nutraceuticals have evolved from the recognition of the relationship between food and health.

The term "nutraceuticals" was devised from combination of "nutrition" and "pharmaceutical" in 1989 by Stephen Defelice MD, founder and chairman of the foundation for innovation in medicine (FIM) Cranford, NewJersy. According to Defelice "nutracetuicals" can be well-defined as a food or part of a food that delivers both medical or health benefits together with the prevention and/or treatment of a disease. The word "Nutraceutical" is made up of two words: - 'nutrients' symbolizing a healthy or nutritious diet and a 'pharmacist' symbolizing a medicinal substance or a substitute for a medicinal substance. The term is used primarily to describe this nutritional or health-related diet that offers health and medical benefits, including prevention and treatment of diseases. These products have extensive variability of products ranging from isolated nutrients, dietary supplements, certain herbs and certain foods to genetically modified foods, herbal products and packaged foods such as grains, soups and drinks (Beisalski, 2001). Nutraceuticals, which are also referred to as medicinal foods, synthetic foods, phyto chemicals, functional foods, emerging foods and nutritional supplements include everyday products such as bio- yoghurts, probiotic milk and whole-grain breakfast cereals, and vitamins, herbs and even genetically modified foods and supplements. It is a alimentary diet supplement that plays an important role in maintaining a healthy human body and provides the necessary supplements for various metabolic processes to regulate body function and, consequently, provide more nutritional supplements and protect the body from disease (Stauffer, 1999). There is a great cornucopia of herbs and foods that promote the support and nourishment of our immune system. Some have been used by the different traditional procedures of several countries and are now being studied by modern researchers.

Nutraceutical Benefits of Food Byproducts

Food waste is produced during all the phases of food life cycle including agricultural production, industrial manufacturing, processing,storage and distribution (Kumar et al. 2017). Following is the list of food byproducts from various food groups explained briefly:

Table 1: Enlisting of food waste derived bioactive components with its potential nutraceutical benefits

Food group	Parts as Byproduct	Bioactive constituent	Potential nutraceutical benefits	Vale added product
Cereals Rice, wheat, barley, ragi, millets	Bran	Γ-Oryzanol, , B-Glucan, Phenolic acids, antioxidants	Antioxidants, antimicrobial,Cholesterol lowering, reducing risk of cvds	Bran oil enriched with omega3 and omega 6 fatty acids
Pulses	Husk	Crude fiber, Carbohydrates		Metal adsorbent
Dairy	Whey	Immunoglobulins (Igs), lactoferrin (Lf), lactoperoxidase (Lp), Glycomacropeptide (GMP) and sphingolipids	Antimicrobial and antiviral properties Anticancer, prebiotic, help in boosting glutathione (GSH) levels	Health drink
Meat fish and poultry	Egg sell	Calcium	Bones and teeth maintenance, bone strength in nerve, muscle and glandular functions.	Edible coating
	Brains, nervous Systems and spinal cords	Cholesterol,	After extraction are used for the synthesis of vitamin D3	Nutritional health supplements milk fortified by cholecalciferol Widely used in Vitamin D deficiency Extraskeletal benefits
Fruits and vegetable	Peaches seeds	Pectin and cellulose	Gel forming activity, prebiotic effects	Jam jelly, marmalade
	Guava grid and seed	Dietary fibre	Laxatives, good bowel movements, lowers high blood pressure relieves constipation reduces prostate cancer risk, helps control cholesterol	Essential oils, food ingredients, medicine and nano-material synthesis
	Mango peel and kernel	Essential amino acids, polyphenols, phytosterols, and tocopherols, oleic and Linoleic acids	Antioxidants, antimicrobial Mango stone as an adsorbent.	Seed kernel oil

Food group	Parts as Byproduct	Bioactive constituent	Potential nutraceutical benefits	Vale added product
	Corn silk (Corn hairs)	Ash, Protein, and dietary fibers	Useful in treating urinary infection and cystitis, reduce blood pressure, cholesterol, blood sugar, and inflammation	As a tea or extract. It may also be taken as a pill.
	Pineapple by-products (residual peels, pulp, stem, stalk and leaves)	Bromelain, sugars, polyphenols, enzymes, organic acids, vitamins, and dietary fibres	Antiedematous, anti-inflammatory, antithrombotic, fibrinolytic activity anticancer agent. Proteolytic activity, tooth whitening	Meat tenderizer, dough Improver, anti-browning agent, beer clarifier, animal feed, cosmetic substance and Textile industry.
	Potato peel	Fibers, phenolic Glycoalkaloids	Antibiotics, anti-allergens, antipyretics, anti-inflammatory and hyperglycaemic, inhibitory activity toward lipid peroxidation	Biodiesel and biomanure, biogas (methane), lactic acid, glycoalkaloids, phenols and biosorbent
	Green peas Peel	Dietary fibers, protein, cellulose, Phenolic	Increase the perception of gastric flow, hypocholesterolaemic properties, reduces starch digestion and glucose distribution	Added to dairy products, bakery products, confectionary and soft drinks.
	Solid waste f carrot juice extraction	Carotenoids, fibers and phenolics	Induce Apoptosis in prostate cancer cells and alignant lymphoblast cells by DNA fragmentation, poly ADP ribose polymerase (PARP) cleavage, and Caspase-3activation	Dairy Products, bakery products, confectionary
	Tomato-seed	Tomatine	Antibiotic, inhibit carcinogenesis in tumour cells.	Tomato-seed oil
	Tomato-peel	Licopen and oligosaccharides		Paste
	Coconut shell, husk	Lignin and cellulose	-	Charcoal
	Jack fruit waste (perianth meal, rind and core meal)	Starch and protein fractions	Antioxidant, antidiabetic, antimicrobial activity	Bakery products, as a cattle feed, extraction of bio-fuel, nano-porous adsorbent for removing dye etc.

Food Byproduct Utilization as Value Added Product

Now a days the demand is shifted to develop nutritious food of natural origin among health-conscious people (Mishra et al. 2022). The plant foods by-products prove to a potential ingredient to develop such enriched product as these are having bioactive compounds such as phenolics, antioxidant, antimicrobial and essential compounds from food industry waste (Charu et al. 2017) which could be utilized as sustainable source to increase the shelf stability of foods. Previous studies have verified that the agro-horticultural wastes contain valuable proportion of phytochemicals and nutrients which may assist as a probable source of nutraceuticals/functional foods. Abundant researches pointed towards extraction of bioactive components such as phytochemicals, phenols and polyphenols (flavonoids, isoflavones, flavonones, tannins, lignins), pigments (carotenoids, phycocyanin, anthocyanins) isoprenoid compounds (tocopherols, tocotrienols, terpenes terpenoids, saponins), carbohydrate (oligosaccharides, non-starch polysaccharides, prebiotics), fatty acid and structural lipids (sphingolipids, essential fatty acids, lecithins), derivatives of amino acid, bioactive peptides, microbes (Lactic acid bacteria, bifiodobacteria and probiotics) vitamins (ascorbic acid) and minerals, natural pigments, dietary-fiber, pectin, enzymes, antioxidants and antibiotics (Sharma et al. 2014) through different extraction technique.

These food byproduct derived bioactive components play vital role for well-being of human health by means of antioxidants, antimicrobial, antifungal, anti-inflammatory, antiallergic, antispasmodic, anticarcinogenic, hepato-protective, chemo-preventive, neuro-protective, hypolipidemic, hypotensive agents, consequentially help in prevention of aging, diabetes mellitus, cardio-vascular diseases, osteoporosis and cancer, induce programmed cell death, diuretic, Central Nervous System (CNS) stimulant, analgesic, immuno-modulator, carminative and protects from carcinogenesis induced by UV-B radiation (Gupta and Prakash 2014). The formulation of new products by utilizing bioactive components with nutraceutical benefits from food waste has been shown in figure 1.

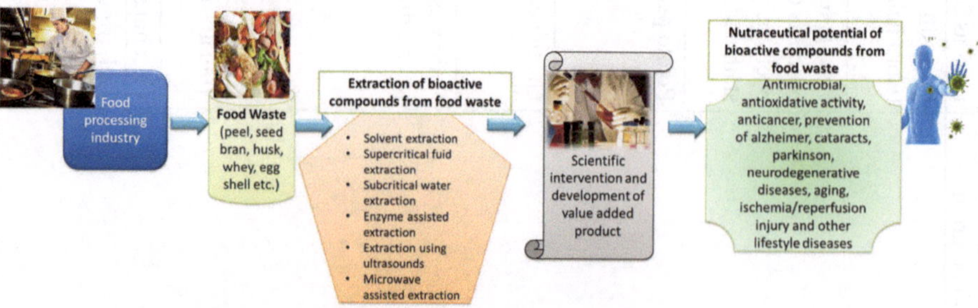

Fig. 1: Utilization of food byproducts into novel industrial products

From the perspective of consumers, functional foods and nutraceuticals food can offer many benefits:

1. May it increase health value according to the healthy supplement of our diet.
2. May it help live longer.
3. May help avoid certain medical conditions.
4. We may benefit mentally by doing something for ourselves and others.
5. It may appear more "natural" than traditional medicine that may have serious side effects.
6. May be provided with food for targeted areas with special needs (e.g. nutrient rich foods for the elderly).

Conclusion

Throughout industrial processing of food, huge amounts of wastes are produced. This has emerged as a thoughtful problematic situation as they employ a major impact on environment which should be alternatively manageable and/or exploited in some other beneficial form. More over utilization of the food processing by-products as functional ingredients sources and plausible applications has developed as a promising field and global necessity payable to improved concern to the environmental safety. Fruits and vegetable processing industries produce huge waste such as peels, seeds, liquid, and molasses as rich source of carbohydrates, proteins, fibers, vitamins, and minerals. Utilization of this waste not only valorizes waste disposal problems but also eliminate environmental pollution. Functional bioactive compounds of natural origin like fruit processing wastes could be utilized as substitute for synthetic additives with multifunctional perceptions by merging wellbeing assistances to technological use. Innovative technical and potential alternative scientific technologies must be used at optimum levels for the extraction of bioactive compounds in addition other compounds of commercial importance from the food wastes. The collective exertion of waste reduction for its sustainable consumption in new product development would noticeably result in reduced quantities of food wastes accumulated globally and help in environmental safety.

References

Biesalski HK (2001), Nutraceuticals: the link between nutrition and medicine. In: Kramer K, Hoppe PP, Packer L, editors. Nutraceuticals in health and disease prevention. New York: MarcelDekker Inc. 1-26.

Charu G, Prakash D, Nazareno MA (2017) Nutraceutical Potential of Agri-Horticultural Wastes. Obes Control Ther 4(2): 1-8.

Gupta C, Prakash D. Phytonutrients as therapeutic agents. J Complement Integr Med. 2014;11(3):151-169.

Lipinski B, Hanson C, Lomax J, Kitinoja L, Waite R, Searchinger T. (2013). Reducing Food Loss and Waste. World Resources Institute, Washington, DC.

Mishra P, Gupta E, Gupta K. (2022) Response surface methodology and textural profile analysis for optimization of fruit peel-based extruded snack (elbow macaroni). J Food Quality, vol., Article ID 5881459, 8 pages https://doi.org/10.1155/2022/5881459

Priedniece V., Spalvins K., Ivanovs K., Pubule J., Blumberga D. (2017) Bioproducts from Potatoes. A Review Environmental and Climate Technologies 21: 18–27

Schieber A, Stintzing FC, Carla R. (2001) By –products of plant food processing as a source of functional compounds-Recent developments. Trends in Food Science and Technology.12(11):401-413.

Sharma G, Prakash D, Gupta C. Phytochemicals of Nutraceutical Importance: Do they defend against diseases? CABI International Publishers, UK; 2014. p.1-19.

Stauffer JE (1999), Nutaceuticals. Cereal Food World, 44(2), 115-116. Kumar K., Yadav A. N., Kumar V., Vyas P. and Dhaliwa H. S. Food waste: a potential bioresource for extraction of nutraceuticals and bioactive compounds. Bioresour. Bioprocess. (2017) 4:18

11

Pulses
and Legume Based Fermented Foods

Pinki Saini[1] and Suman Devi[2]

[1]Centre of Food Technology, University of Allahabad, Prayagraj, Uttar Pradesh
[2]College of Dairy Technology (Etawah Campus), Chandra Shekhar Azad University of Agriculture & Technology, Kanpur, Uttar Pradesh

Abstract

Fermentation in food processing is the process of converting carbohydrates to alcohol or organic acids using microorganisms; yeasts or bacteria under anaerobic conditions. Fermentation usually defines the desirable action of microorganism to enhance food safety by inhibition of pathogens, modify their sensory properties, improve their nutritional value by removing anti-nutritive compounds and by enhancing bioavailability of components. Pulses and Legumes are the major source of dietary nutrients all over the world. Fermentation can be the most simple and economical way of improving their nutritional value, sensory properties and functional qualities. Pulses and legume based fermented food products offer opportunities to include probiotics, prebiotics and fibers in the human diet. In the Indian subcontinent, making use of fermented food and beverages using local food crops and other biological resources is very common, although the nature of the products and the base material varies in different regions. This chapter focuses on some of the indigenous fermented foods produced in the country that have not received the scientific attention they deserve in the last decades. Products produced from different pulses and legumes substrates fermented by lactic acid bacteria, yeast and/or fungi are included.

Introduction

Since the beginning of human civilization there has been an intimate companionship between the human being and the fermentative activities of microorganisms. These fermentative activities have been utilized in the production of fermented foods and beverages, which are defined as those products that have been subordinated to the effect of microorganisms or enzymes to cause desirable biochemical changes (Harlander, 1992). Fermentation is one of the oldest and most economical methods of producing and preserving food (Billings, 1998; Chavan & Kadam, 1989).

Fermented pulse products have been an indigenous part of the cuisines of East and Southeast Asia. The preparation of most of these traditional fermented foods and beverages till today, remains as a household art. They are generally produced in homes, villages and small-scale industries whereas on the contrary, the preparation of some products, such as soy sauce, tempeh etc. has evolved to a biotechnological state and is carried out on a large commercial scale (Bol & de Vos, 1997). In the distant past, there was no verified data on the economic, nutritional, technical and quality control implications of the indigenous fermented food. However, in the last few decades, the number of studies dealing with indigenous fermented beverages and foods around the whole world has rapidly increased (Steinkraus et al., 1993).

In this context, this chapter aims to list and summarize the production processes of some of the most common traditional pulses and legume based fermented foods. In addition, some of the microbiological and nutritional properties of those foods and the technological improvements which have been achieved on its production are being highlighted.

Pulses

Pulses play an important role in providing a nutritionally balanced diet. These are the principal source of protein for vegetarians. The total world acreage under pulses is about 93.18 (Mha) with production of 89.82 (Mt) at 964 kg/ha yields level. India, with >28 Mha pulses cultivation area, is the largest pulse producing country in the world. It ranks first in area and production with 31 per cent and 28 per cent respectively (dpd.gov.in., 2022).

The cultivation of pulses has gradually been pushed to non-irrigated, marginal lands. Unlike other major crops, the pulse crop is risk prone, being vulnerable to pest attacks and sensitive to weather. Both high as well as a low rainfall affect the pulse crop adversely. Farmers have little hope of getting remunerative prices, even when consumer prices are high (http://agropedia.iitk.ac.in). Out of the total area >60 Lha is confined to Madhya Pradesh alone, earning a prime status in pulse production commodity contributing a remarkable 21% of the country's pulse area with 25% production, thereby ranking first both in area and production followed by Rajasthan, Maharashtra and Uttar Pradesh with 16%, 15% and 10%. ß More than 90 per cent of total pulse production has been contributed by 10 states of Madhya Pradesh, Rajasthan, Maharashtra,

Uttar Pradesh, Karnataka, Andhra Pradesh, Gujarat, Jharkhand, Tamilnadu and Odisha (dpd.gov.in., 2022).

Nutritive quality of pulses

Pulses provide protein, complex carbohydrates, and several vitamins and minerals. Like other plant-based foods, they contain no cholesterol and little fat or sodium. Pulses also provide iron, magnesium, phosphorus, zinc and other minerals, which play a variety of roles in maintaining good health. The protein content of pulses are twice that of cereals (20 - 25%) and almost equal to that of meat and poultry. But the quality of protein content is inferior to animal protein. Pulses are lacking in amino acid methionine but have higher lysine content. This limitation can be overcome by combining pulses with cereal. Pulses contain negligible amount of fat with exception of soya bean (19.5%). Unsaturated fats are the major types of fat present in pulses. Among the vitamins, pulses contain fair source of thiamine, riboflavin and nicotinic acid. The iron content is high in pulses, about 8- 10 mg per 100 g. Pulses and legumes also contain potassium, phosphorous, manganese and magnesium. Whole pulses contain insoluble and soluble fibre. Insoluble fibre guards against constipation, so reduces the risk of the colon and rectum cancer whereas soluble fibre helps in lowcring blood cholesterol levels and thus reduces the risk of cardiovascular diseases. The starches and carbohydrates content of pulses are complex type and are digested and absorbed slowly. This helps to a steady release of glucose into the blood and is useful for diabetics to control their blood sugar levels (Case and Fenster, 2011). Soya beans, because of its phytoestrogen content, believed to have a role in protecting breast cancer, osteoporosis and menopausal symptoms. As a result of their nutrient content and other properties, pulses can play a role in several special diets:

- **Gluten-free diet:** If a person with celiac disease consumes gluten (a protein found in wheat and some other cereal grains), an immune reaction is triggered in the small intestine, which can cause damage and poor absorption of nutrients. Pulses contain no gluten; therefore, people with celiac disease can use chickpeas, lentils or peas as an ingredient in recipes (Case and Fenster, 2011).

- **Diabetic diet:** For people with diabetes, consuming lentils, peas and beans may help with blood glucose management. Compared with some other carbohydrate sources, pulses have a lower glycemic index. Some studies have shown that consuming pulses may result in more stable blood glucose levels after meals.

- **Vegetarian diet:** Pulses are good sources of protein, vitamins and minerals (especially iron and zinc), which makes them an excellent food choice for vegetarians. They contain eight essential amino acids. Consuming lentils with rice provides the full complement of amino acids needed for growth.
- **Weight management diet:** Although more studies are needed, consuming pulses may help with weight management. For people trying to lose weight, pulses are high in fiber and protein, low in fat and moderate in calories. One cup of cooked lentils or dry peas contains about half of the daily fiber recommendation for adults. Foods higher in fiber content usually help people feel "full" or satiated at mealtime.

Fermentation

Fermentation may be defined as any process for the production of a product by the mass culture of microorganism (Stanbury, 1999). The microorganisms responsible for the fermentation may be the microflora indigenously present on the substrate, or they may be added as starter cultures (Harlander, 1992). Fermentation provides a natural way to reduce the volume of the material to be transported, to enhance the nutritive value and appearance of the food, to reduce the energy required for cooking and to make a safer product (Simango, 1997). Indigenous people have been using microbes unknowingly for various purposes (Sekar and Mariappan, 2007). Moreover, fermented foods have further benefits of providing bio-nutrients and minerals and also used for enhancement of flavour and aroma. The process also increases digestibility and exert health promoting benefits (Jeyaram et al., 2009). Fermentation may assist in the destruction or detoxification of certain undesirable compounds (phytates, polyphenols and tannins) which may be present in raw foods (Sharma and Kapoor, 1996).The traditional way of carrying out fermentation at the household-scale is still followed using relatively simple processing facilities. These products often contain mixed microbial populations because of the lack of sterility and the use of natural fermentation (Nout and Sarkar, 1999).

Fermented foods are produced using various manufacturing techniques, raw materials and microorganisms. However, there are only four main fermentation processes: alcoholic, lactic acid, acetic acid and alkali fermentation (Soni and Sandhu, 1990). Alcohol fermentation results in the production of ethanol, and yeasts are the predominant organisms (e.g. wines and beers). Lactic acid fermentation (e.g. fermented milks and cereals) is mainly carried out by lactic acid bacteria. A second group of bacteria of importance in food fermentations are the acetic acid producers from the *Acetobacter* species. *Acetobacter* convert alcohol to acetic acid in the presence of excess oxygen. Alkali fermentation often takes place during the fermentation of fish and seeds, popularly used as condiment (McKay and Baldwin, 1990).

Nutritional value of fermented foods

Generally, a significant increase in the soluble fraction of a food is observed during fermentation. The quantity as well as quality of the food proteins as expressed by biological value and the content of water soluble vitamins is generally increased, while the antinutritional factors show a decline during fermentation (Paredes-López and Harry, 1988). Fermentation results in a lower proportion of dry matter in the food and the concentrations of vitamins, minerals and protein appear to increase when measured on a dry weight basis (Adams, 1990). Single as well as mixed culture fermentation of pearl millet flour with yeast and *lactobacilli* significantly increased the total amount of soluble sugars, reducing and non-reducing sugar content, with a simultaneous decrease in its starch content (Khetarpaul and Chauhan, 1990). Combination of cooking and fermentation improved the nutrient quality of all tested sorghum seeds and reduced the content of antinutritional factors to a safe level in comparison with other methods of processing (Obizoba and Atii, 1991). Mixed culture fermentation of pearl millet flour with *Saccharomyces diastaticus, Saccharomyces cerevisiae, Lactobacillus brevis* and *Lactobacillus fermentum* was found to improve its biological utilisation in rats (Khetarpaul and Chauhan, 1991). Fermentation induced a significant decrease in lipid and lignin contents of okara, which is an insoluble residue obtained as a by-product in the manufacture of soybean milk. The fermented okara on the other hand neither increased PER nor the weight gain in rats (Guermani et al., 1992) compared to non-fermented samples. The digestibility of starch in bengal gram, cowpea and green gram was increased by fermentation. Cooking of these fermented legumes further increases the starch digestibility (Urooj & Puttaraj, 1994).

The germination of pulses increases its nutritional content. The germinated pulses have high vitamin C content than the normal seed (Tang et al., 2014). Other vitamins that show an increase are B complex vitamins and folic acid. Tannin and phytates, those adversely affect the bioavailability of nutrients, are broken down on germination. The chlorophyll content also increases when the sprouted shoots change from yellow to green.

Health effects of fermented foods

One of the reasons for the increasing interest in fermented foods is its ability to promote the functions of the human digestive system in a number of positive ways. This particular contribution is called probiotic effect. Already early in 1900, Metchnikoff pointed out the use of fermented milks in the diet for prevention of certain diseases of the gastrointestinal tract and promotion of healthy day to day life. Since then a number of studies have now shown that the fermented food products do have a positive effect on health status in many ways. The human intestinal microbial flora is estimated to weigh about 1000 grams and may contain 10^{16}– 10^{17} colony forming units representing more than 500 strains. For physiological purposes, it can be considered to be a specialized organ of the body with a wide variety of functions in nutrition, immunology and metabolism (Gustafsson, 1983). Studies on mice have shown that the indigenous microorganisms in the stomach are *Lactobacillus, Streptococcus* and *Torulopsis*, while in the small intestine, ceacum and colon several different species (*Bacteroides, Fusobacterium, Eubacterium, Clostridium,* etc.) coexist (Savage, 1983). The gastrointestinal microfloras in humans are also known to contain hundreds of species. Even though there is a wide variation among individuals, the number of species and size of the population are usually kept stable in normal healthy subjects. There is a constant struggle in maintaining the desirable balance and a dynamic equilibrium between microbial populations within the intestinal flora (Robinson and Samona, 1992). The stability of the intestinal microflora is affected by many factors including dietary habits. Under normal conditions the intestinal wall prevents translocation of organisms both dead and living as well as microbial products like toxins from the gut to the blood. However, in patients with systemic insult like starvation, shock, injury and infection or specific insult of the gastrointestinal canal through inflammation, chemotherapy or radiation, the gut mucosal permeability will be increased leading to translocation of microbes (Carrico and Meakin, 1986; Alexander et al., 1990; Wells, 1990; Kasravi et al., 1997). A fermented food product or live microbial food supplement which has beneficial effects on the host by improving intestinal microbial balance is generally understood to have probiotic effect (Fuller, 1989).

Apart from this, there are interesting data on anticarcinogenic effect of fermented foods showing potential role of *lactobacilli* in reducing or eliminating procarcinogens and carcinogens in the alimentary canal (Reddy et al., 1983; Shahani, 1983; Mital & Garg, 1995). The enzymes α-glucuronidase, azoreductase and nitroreductase, which are present in the intestinal canal, are known to convert procarcinogens to carcinogens (Goldin and Gorbach, 1984). Oral administration of *Lactobacillus rhamnosus* GG was shown to lower the faecal concentration of α-glucuronidase in humans (Salminen et al., 1993) implying a decrease in the conversion of procarcinogens to carcinogens. The process of fermentation of foods is also reported to reduce the mutagenicity of foods by degrading the mutagenic substances during the process. However, a review by McIntosh (1996) concludes that there is only limited data to support the hypothesis that probiotic bacteria are effective in cancer prevention. On the other hand, a study by Hosono and Hisamatsu (1995) on the ability of the probiotic bacteria to bind cancerogenic substances have reported that *E feacalis* was able to bind aflatoxin B_1, B_2, G_1 and G_2 as well as some pyrolytic products of tryptophan. During fermentation of the beans for preparation of *tempeh*, the trypsin inhibitor is inactivated, and the amount of several oligosacharides which usually cause flatulence was significantly reduced (Hesseltine, 1983). Bean flour inoculated with *Lactobacillus* and fermented with 20% moisture content, showed a reduction of the stachyose content in the food (Duszkiewicz-Reinhard et al., 1994)

Pulse based fermented foods

Fermented foods made from legumes constitute an important part of the human diet in many developing countries, including India (Sandhu and Soni, 1989) (Table 6.1). Fermented soybean products have been reported to be used extensively in almost all the states of North-East India and bears resemblance to *tou-shi, hamanatto, chiang-yu, shi-tche, chiang* and *tofu* of China, *tempe kedele, kecap* and *taoco* of Indonesia, *shoyu* and *miso* of Japan (Wang and Fang, 1986; Nout, 1995). The whole grains contain flavonoids, terpenoids and other natural antioxidants like carotene, ascorbic acid and tocopherol. An increase in the content of crude protein and a decrease in fat content have been observed in some traditionally fermented bean product. Also, a significant decrease in crude fibre content has been observed (Gupta and Nagar, 2010). The high crude protein content may be attributed to the microflora developed during the fermentation process (Gupta et al., 2007).

Idli

A fermented, thick suspension made of a blend of rice (*Oryza sativum*) and dehulled black gram (*Phaseolus mungo*) is used in several traditional foods in Southeast Asian countries. Among them, *idli* and *dosa* are very popular in India and Sri Lanka (Sands & Hankin, 1974). Traditionally, for idli preparation the rice and black gram are soaked separately. After draining the water, rice and black gram are grinded independently, the rice is coarsely ground and the black gram is finely ground. Then the rice and the black gram batters are mixed together (2:1ratio) with addition of a little salt and allowed to ferment overnight at room temperature. Finally, the fermented batter is placed in special *idli* pans and steamed for 5–8 min (Nagaraju & Manohar, 2000).The lactic acid bacteria i.e. *Lactobacillus delbrueckii, Lactobacillus fermenti, Lactobacillus lactis as well as Leuconostoc mesenteroides, Streptococcus faecalis, and*

Pediococcus cerevisiae have been found to be responsible for the fermentation process, although *L. mesenteroides* and *S. faecalis* are considered to be the microorganism essential for leavening of the batter and for acid production in idli (Purushothaman, Dhanapal, & Rangaswami,1993; Ramakrishnan, 1993). The yeasts *Geotrichum candidum, Torulopsis holmii, Torulopsis candida* and *Trichosporon pullulans* have also been identified in *idli* fermentation (Chavan & Kadam, 1989; Shortt, 1998).Fermentation of *idli* batter appears to have a significant effect on the increase of all essential amino acids and in the reduction of antinutrients (such as phytic acid),enzyme inhibitors and sugars (Steinkraus et al.,1993).

Idli is a low calorie, starchy and nutritious food, which is consumed as breakfast or snack. Steamed *idli* contains about 3.4% protein, 20.3% carbohydrate and 70% moisture (Teniola & Odunfa, 2001). Large-scale production of *idli* is carried out in batch compartmental steaming units. This is labour intensive and has limited capacity. With the growing demands for breakfast foods, *idlis* are being consumed on a large scale in some Indian institutions such as army, railways, industrial canteens, etc. In order to meet the demand, numerous studies are carried out for the development of continuous units for the production of *idli* (Murthy, Nagaraju, Rao & Subba Rao, 1994; Murthy & Rao, 1997; Nagaraju & Manohar, 2000).

Dosa

It is very similar to *idli* batter except that the rice and black gram are finely ground and that the fermented suspension instead of being steamed is heated with a little oil, on a flat plate. A *dosa* suspension is prepared by grinding wet rice and black gram separately with water. The two suspensions are then mixed and allowed to undergo natural fermentation, usually for 8–20 h. To make a dosa, the fermented suspension is spread in a thin layer (of 1–5 mm thickness) on a flat heated plate, which is smeared with a little oil or fat. A sol to gel transformation occurs during the heating and within a few minutes, a circular, semi-soft to crisp product resembling a pancake, ready for consumption is obtained (Battacharya & Bhat, 1997). The microbiological, physical and biochemical changes of *dosa* during fermentation and its nutritive value are quite similar to *idli* (Chavan & Kadam, 1989; Purushothaman et al., 1993; Ramakrishnan, 1993; Sands & Hankin, 1974; Shortt, 1998).

Soy sauce

Soy sauce is a dark brown liquid, made from a blend of soybeans and wheat that is mainly used as an all purpose seasoning in Japan, China and the Far East countries (Yokotsuka, 1993). Soy sauces have a salty taste, but are lower in sodium than traditional table salt. The traditional manufacturing techniques for soy sauce have been well described (Beuchat, 1983; Xu, 1990; Yokotsuka & Sasaki, 1998). Cooked soybeans are mixed with coarse wheat flour, with adjustment of the initial moisture of the mixture to about 55% (w/w). The soybean–wheat mixture is inoculated by molds, and after 3 days of fermentation at 25–35 _C, the soybeans and flour mixture (known as *koji*) is immersed in a brine solution (22–25%), the ratio of *koji* to brine is about 1:3 (w/v). After mixing, the salt concentration of the mixture usually falls to about 18–21%.

This brine solution containing *koji* is known as *moromi*. The *moromi* is left to ferment for a period of 1–12 months; the longer the fermentation time the better the soy sauce quality. After the *moromi* fermentation is completed, the liquid part (soy sauce) is separated, filtered, pasteurised and bottled (Ayres et al., 1993; Franta et al., 1993; Mensah, 1997; Rowan et al., 1998; Steinkraus, 1998). There have been many studies describing the involvement of various microorganisms in soy sauce fermentation and three major groups have been reported in the literature (Horitsu et al., 1991; Scheinbach, 1998; Yeoh, 1995; Yong and Wood, 1974, 1976, 1977). These include fungi such as *Aspergillus oryzae* and *A.soyae* involved in the *koji* production, halotolerant LAB and yeast strains such as *Zygosaccharomyces rouxii* and Candida species responsible for the *moromi* fermentation (Hamada et al., 1991; Allen et al., 1996; Roling et al., 1996; Holzapfel et al., 1998).

The characteristic aroma and flavour of soy sauce is due to the enzymatic activities of yeasts and some LAB. As soybeans contain high levels of proteins and oligosaccharides, but lack in simple sugars, fermentation by lactic acid bacteria and yeasts requires the exogenous saccharifying enzymes supplied by the *koji*. In general the pH of the sauce is between 4.6 and 4.8, and the typical salt concentration is 17–19%. Concentration of salt less than 16% can result in the development of putrefactive species during fermentation and ageing. On the contrary, levels greater than 19% interfere with the growth of halophilic bacteria such as *Pediococcus halophilus* and osmotic yeasts such as *Z. rouxii* (Beuchat, 1983). During the last 2–3 decades the traditional process for soy sauce production has been significantly improved. Some examples of these developments include the introduction of improved microbial inocula for soy sauce fermentation, the use of new techniques for preparing koji, and the use of new materials and modern technology for processing and fermentation (Mensah, 1997).

Tempeh

Tempeh is a fermented soyfood and is unique in its texture, flavor and versatility. It originated in Indonesia, where today it is still the most popular soyfood (Winaro and Reddy, 1986). *Tempeh*, while not as popular as tofu in the United States, lends itself easily to being used as a meat alternative because of its chewy texture and distinct flavour. As a result, a wide variety of *tempeh*-based meat analogues are available. *Tempeh* is a cake of cooked and fermented soybeans held together by the mycelium of *Rhizopus oligosporus*. The soybeans are dehulled mechanically or by soaking. Soaking is generally done at 100°C for 30 min in acidified water (lactic or acetic acid is used at pH 4.3 to 5.3). Cooking is done in the same water at 100°C for 90 min. The water is drained and the soybeans are dried, inoculated and incubated at 35-38°C for 18-24 hrs. Storage is done at refrigeration temperature; it can then be further processed or packaged and pasteurized for distribution.

Miso

Miso, a white, brown or reddish-brown soybean paste, is another fermented soyfoods product. Made from fermented soybeans, and sometimes in combination with wheat, barley or rice, this salty paste is an important soup base and flavoring ingredient used throughout Japan, Korea, Taiwan, Indonesia and China (Ebine, 1986). There are many types of *miso* available in the United States today, from sweet white *miso* to the more savory dark *miso*. *Miso*, sold extensively in natural food stores and Oriental markets, has developed a following among New Age cooks and health enthusiasts who savor its rich flavours and believe in its purported healthful benefits. To produce *miso*, whole soybeans are soaked in water and cooked until tender. The cooked soybeans are dusted with *Aspergillus oryzae* a fungal starter and shaped into *koji* nuggets. This can be done in combination with either rice or barley. These nuggets are incubated, during which the fungus matures on the soybeans causing them to become white and fuzzy. Enzymes and nutrients produced during this stage are essential to later stages of fermentation. The mature nuggets are mixed with salt and water in fermentation vats, resulting in the formation of a paste. During aging, the flavors and aromas are produced. When the *miso* is fully ripened, it is blended, pressed and packaged (Ebine, 1986).

Natto

It is a traditional Japanese food made by boiling or steaming soybeans and fermenting them with the *Bacillus subtilis natto*. The *Bacillus subtilis natto* is a rod-shaped bacterium with a relatively high heat tolerance (40°–50°C) compared to other bacteria. The bacillus bacteria activate a fermenting process that provides the beans their exclusive sour aroma, nutty flavour, and slippery texture. *Natto* possesses an odour of ammonia as a result of the breakdown of amino acids. During the *natto* fermentation viscous and sticky polymers (polyglutamic acid) are produced which makes the *natto* very slimy. During the fermentation the bacteria produces enzymes, vitamins, amino acids and other nutrients unique to *natto*. Some of the beneficial effects of *natto* include prevention of heart attacks, cancer, osteoporosis and intestinal infections. *Natto* bacteria have a strong enzyme, which can dissolve blood clots. It has been found that *natto* contained a potent fibrinolytic enzyme named as nattokinase (Sumi et al., 1987).*Natto*kinase, extracted from *natto*, is a fibrinolytic enzyme that effectively breaks down fibrin strands and the thrombi this fibrin holds together. Nattokinase has a protein structure made up of a single polypeptide chain consisting of 275 residues of Ala at the N-terminal. It strongly hydrolyzes fibrin. Nattokinase is the most potent fibrinolytic enzyme found among the approximately 200 foods that have been investigated for oral fibrinolytic therapy (Sumi, 1998)

While some researchers have theorized that *natto* may have prehistoric origins, it first became a part of the Japanese culture during the later part of the Edo Period (1600–1868). During this period, soybeans were packed in straw and buried underground for a week or more. A naturally occurring bacillus in the straw facilitated the soybean fermentation. The modern method of making *natto*, by injecting the *Bacillus*, was first developed in Sendai. In controlled trials, *natto* is prepared by adding *Bacillus subtilis natto* to steamed soybeans in polystyrene containers and then allowing the mixture to ferment for 24 hours at37°C. The *natto* bacilli are used in a common delivered concentration of 10^{10} cells/g of dried bacilli powder (Sumi, 1998).

Kinema

It is a pulse based fermented food made of yellow seeded soybeans in contrast to brown or black seeded ones. It has originated in Darjeeling and Sikkim and considered as a low cost source of high protein flavoured food.

Soybeans are washed and soaked in water over night (12-20 hours) at ambient temperature (10-25°C). The beans increase about 2-3 times the original weight and the pH of soaked water become acidic (~5). The beans are drained and cooked by boiling (90-95°C) in fresh water until they can easily crushed between the finger tips (usually 90 min of boiling). The beans are drained and crushed lightly by a wooden pestle to make grits of mainly half cotyledons. A small amount of fire wood ash is often added. The grits are wrapped usually in fern leaves (*Athyrium* sp.) or *Leucosceptrum canum*, covered with sack cloth and kept in a bamboo basket above an earthen oven in kitchen at ambient temperature to ferment for 1-3 days. Instead of fern leaves, *Ficus* and banana leaves can also be used as a wrapping material (Sarkar et al., 1993; Sarkar and Tamang 1994; Sarkar et al., 1996; Tamang and Nikkuni 1998).

The desired state of fermentation determined primarily by a typical *kinema* flavour (having a pungent smell of ammonia) as well as a white viscous fluid on the beans. The fluid has the property of forming long, string threads when touched with finger the longer the strings, the better the quality of *kinema*. Fresh *kinema* of good quality has a nutty flavour accompanied by a mild smell of ammonia, a greenish brown color and is semi hard like raisins. *Bacillus subtilis* is the predominant microorganism involved in the production of kinema. The other microorganisms that are present are *B. licheniformis, B. circulans, B. thuringiensis,B. sphaericus*, *Enterococcus faccium*, *Candida parasilosis* and *Geotrichum candidum.* The surface to the fermenting soybean changes to a whitish colour due to growth of *B. subtilis* on or between the soybeans during fermentation. Stickiness is an important criterion for judging quality of *kinema* by consumers. Water soluble nitrogen and formal nitrogen to total nitrogen contents of *kinema* increases rapidly during fermentation. This was due to high proteolytic activity of *B. subtilis*. Reducing sugar increased at log phase and then decreased sharply during *kinema* fermentation. This indicates that reducing sugars of fermenting soybeans were used by *B. subtilis* for its metabolism (Owens et al., 1997). After soaking soybeans in water, thiamine (B_1) content decreased, whereas riboflavin (B_2) content remained unchanged. Cooking had no influence on the B_1 content, but it enhanced the level of B_2 and niacin (B_3). Incubation of beans at 37°C for 48 hours, when mixed with *Bacillus subtilis*, caused an increase in concentration of both B_1 and B_2. Vitamin B_1 levels decreased when either *Enterococcus faecium* accompanied *B subtilis* or the temperature was elevated for 18 hrs fermentation. (Sarkar et al., 1997; Tamang and Nikkuni, 1998). While the vitamin B_1 content was significantly ($P<0{\cdot}05$) higher, the contents of vitamins B_2 and B_3 were significantly ($P<0{\cdot}05$) lower in raw soybeans than those in *kinema*. Mineral concentrations were 3·1-8·3 times higher in raw soybeans than in *kinema*. The fresh kinema contains total dry matter content of 39.1% (w/w), water soluble dry matter content is 25.6% of total dry matter, pH of 6.25 and free ammonia nitrogen content of 1.19 mM/g of dry matter (Kiers et al., 2000, Sarkar et al., 2002,).

When fresh *kinema* is fried in oil, within first few minutes, the ammoniacal smell vanishes leaving a persistent nutty odor. After a brief frying, *kinema* is added with vegetables, spices and salt to prepare a thick curry. The major desirable aspects of *kinema* are its attractive flavour; certain nutritional qualities and about 50% reduced cooking time compared with soybeans. *Kinema* keeps for 2 to 3 days during summer and a maximum of 1 week in winter. *Kinema* contains 48% (dry weight basis) crude protein, glutamic acid was the most abundant amino acid followed by aspartic acid (Nout et al., 1998 and Sarkar, 2000). The total basic amino acids (ie. lysine, histidine and arginine) constituted 15.1% of the total. Aromatic amino acids and praline accounted for 13.0% and 5.6%, respectively of the total. Tryptophan, cysteine and methionine were the major limiting amino acids in *kinema* (Kiers et al., 2000; Sarkar et al., 2002; Tamang, 2003; Singh et al., 2007).

Tungrymbai

It is also a pulse based high protein food made from soybeans, originated in Meghalaya. Cleaned and washed soybean seeds are soaked in water for about 4-6 hours. The outer skin is removed before cooking by rubbing the soaked seeds gently between the palms. The separated skins are discarded and the seeds are cooked for one hour in the same water till all the water is absorbed. The product is then allowed to cool down after which, it is packed in a bamboo basket lined with lamet leaves (*Clinogyne dichotoma).* The basket is completely covered with lamet leaves and a thick cloth is placed over it to prevent contact with air and maintain warm conditions necessary for fermentation. After proper covering, the basket is kept over the fireplace of the house to maintain warmth. The incubation is done for 3-5 days to get the desired product. The micro-organisms involved in the fermentation are *Bacillus subtilis*, LAB and yeast (Dike and Odunfa, 2003.)

The colour of the well fermented product is brown with a characteristic odour and mild pungency. *Tungrymbai* contains 45.9% of protein, 30.2% of fat, 12.8% of fiber, 5.5% of ash while dry matter is 28.9%. In addition to this, it is found to be rich in various minerals like Ca, P, Fe, Mn, Cu, Zn, Na, K, Mg and Se (Murughar and Subblakshmi, 2006). The carotene and folic acid content are212.7μg and 200 μg, respectively per 100g of product. (Tamang et al., 2009; Sohliya et al., 2009).

Axone

It is also known as *Akhone*, *Kheuha, Sabrocha* and *Sadocha*. It is a soybean based food originated in North eastern India. The beans are boiled to soften and the water is drained out. The cooked beans are then wrapped in leaves of banana or leaves of *Phrynium pubinerve* Blume (Marantaceae) or *Macaranga indica* Wight (Euphorbiaceae) and kept in the bamboo shelf above the fireplace to ferment for a week. When seeds start emitting flavour, they are taken out from the leaves and ground in the wooden pestle called *tathu khuo*. The ground paste is packed in a banana leaf and exposed to sunlight or placed near the fireplace and left for one to two weeks. The beans are used in chutney preparation along with chilly, tomato and salt. The microorganism involved is *Bacillus subtilis*. For long term storage and depending on choice of taste, the fermented beans are kept in cake form above the fireplace or individual beans are separated, dried in the sun and stored in containers. (Singh et al., 2007; Mao and Odyuo, 2007; Tamang et al., 2009.)

Bekang

It is also a high protein soybean based food originated in Mizoram. The seeds of soybeans are crushed and boiled in water for about half an hour and taken out. Boiled seeds are kept in bamboo plates to drain out the water. After some times, the lukewarm seeds are wrapped in leaves of *belphuar* tree (*Trema orientalis* Blume, Family Ulmaceae). Only the leaves of *belphuar* trees are used for maintaining the temperature during fermentation. The wrapped bundle is placed over a bamboo shelf near the fire place and the fermentation is completed within 3 days. Alternatively, small sized soybeans are soaked for 10-12 hours, boiled and wrapped in leaves of *Calliparpa aroria* or leaves of *Phrynium* sp. and kept inside the bamboo basket. It is kept near the earthen oven and fermented for 3-4 days. Sticky beans with ammoniacal flavour are produced to get *bekang*. The microorganism involved is *Bacillus subtilis* (Singh et al., 2007).

It is served after adding desired quantity of chilli and ginger or consumed directly. It is the integral part of all the leafy vegetable based boiled foods. For longer storage life, fermented soybean seeds are spread in the sieve and dried in the sun; dried seeds are packed in bamboo container and closed. After 10-12 days, it is taken out, made into paste and stored in *belphuar* tree leaves (Tamang et al., 2009).

Aagya

It is a pulse based food originated in Arunachal Pradesh made of soybeans. Desired quantities of soybean seeds are boiled in an aluminum container for about 30 minutes. Boiled seeds are kept in oko leaves (*Phrynium pubenerve* Blume, Family Maranphaceae) to drain out water. Then, the seeds are tightly packed in the *Oko* leaves are kept in bamboo shelf ('*rapkho'*) above the fire place in the kitchen for 12-14 days. After this, it is taken out and stored in a basket made of bamboo known as '*udu'*. It is generally consumed mixed with a variety of local vegetables, chutney and boiled foods (Singh et al., 2007)

Peron naming

It is also a soybean based food originated in Arunachal Pradesh. For the preparation of this product soybean seeds are cleaned to remove the chaff and washed. The cleaned seeds are boiled in water for about 30-40 minutes till they become very soft. They are cooled by spreading over bamboo mat ('*epo'*). The seeds are transferred into the leaves of *Phrynium pubenerve* and packed with bamboo or cane rope and hung near the kitchen or kept in the bamboo shelf called *Peron* for fermentation. When the fermentation process is completed after 10-12 days (in summer) or 25-30 days (in winter), the seeds are taken out. These fermented seeds, called *Peron naming* are mixed with several boiled vegetables along with chilli, fermented bamboo shoots and salt. It can also be also used by making paste of fermented soybean seeds mixed with the paste of tomato, chilli, ginger and salt. It can be further processed into paste, made into cake or ball shape and kept in the *Phrynium pubenerve* leaves near the fireplace in kitchen.(Singh et al., 2007).

Yanni perung

Another fermented product originated from Arunachal Pradesh used as a side dish for meals, made of soybeans. The Seeds are cleaned, washed and boiled in water for about 30 min. The lukewarm seeds are placed in the *kulu* leaves (*Musa balbisiana* Colla), packed airtight and kept over fireplace for smoking. The seeds get fermented within 3-4 days during summer while in winter it may take 10-12 days. The fermented seeds are taken out and ground into chutney with tomato, chilli, ginger, garlic and salt. It can be preserved for over several months after drying and making its paste with red chilli and dried ginger (Singh et al., 2007).

Hawaizaar

It is a pulse based food originated in Manipur. It is used as a condiment, most often in the festival *Gobardhanpuja*, *Surap* (ritual ceremony after 13 days of death of a man) and *Asti* (ritual ceremony after 6 days of death of a man). Desired quantities of soybean seeds are taken, washed and properly cleaned. The seeds are then boiled in water for about 2-3 hours till they become soft and excess water is drained out. The lukewarm seeds are wrapped in the cotton cloth and tied airtight with the banana leaves. It is kept inside the bamboo basket (*thumok*) covered with rice husk and pressed with stone. The *thumok* is kept over the fireplace for fermentation, which takes about 4-5 days in summer and 7-10 days in winter. After fermentation, *hawaizaar* is dried in sunlight for 10-15 min and a paste is made after crushing it. The inside of earthen pot (*saphu*) is smeared with mustard oil and paste of *hawaizaar* is kept inside it. The mouth of earthen pot is closed airtight with banana leaves to preserve longer and avoid further fermentation. The *hawaizaar* is used in preparing varieties of chutney and consumed with fermented fish (*ngari*) and salad (Singh et al., 2007).

Peruyyan

It is a soybean based food of Arunachal Pradesh. Soybean is washed with water and cooked till the beans become soft. The excess water is drained off and is cooled for some time. The cooked beans are kept in bamboo basket lined with ginger leaves, locally called as 'taki yannii'. The basket is loosely covered with ginger leaves and is kept on the wooden rack above the fire place for fermentation. Between 3-5 days, the stickiness of the product is checked, and if the product is sticky enough then the product is ready for consumption. The microorganisms involved are *Bacillus subtilis* and Lactic acid bacteria (Tamang et al., 2009).

Soybean (libi) Chhurpi

Local varieties of soybean seeds are boiled and spread on bamboo mat (*charang*) after draining out the water. Salt is added to the seeds when still lukewarm and mixed properly. The drained water is not thrown; it is used for malicing on skin for curing cracks during winter. The drained water can be preserved only for 1-2 days. After cooling, the seeds are put in a container ('*Shong*') and covered with leaves of Zola or banana. After putting the lid, *Shong* is kept near fire and is rotated regularly. When smell comes and seeds become thread like, they are taken out, crushed carefully and spread in small heaps on bamboo mat (*charang*) and sun dried. Again, it is crushed and dried. Crushing is done three times. The prepared *chhurpi* can be stored in dried bottle gourd or bamboo containers for 1-4 years. The place of origin and usage of the product is Arunachal Pradesh (Singh et al., 2007).

Wadi

The place of origin or usage is Punjab and West Bengal. These are dried, hollow, brittle cones or balls (3-8 cm diameter, 15-40g in weight) that are used as a spicy condiment or an adjunct for cooking vegetables, grain legumes or rice. Generally black gram is used for making wadi. The seeds are soaked, drained, ground into smooth soft dough, left to ferment for 1-3 days, molded into cones or balls, deposited on bamboo or palm mats smeared with oil and sun dried for 4-8 hours. The surface of the cones or balls becomes covered with a mucilaginous coating which helps to retain the gas formed during their fermentation. The *wadis* look hollow, with many air pockets and yeast spherules in the interior and a characteristic surface crust. The microorganisms involved are *L. mesenteroides*, *L. fermentum*, *S. cerevisiae* and *Trichosporon cutaneum* become dominant. *Candida vartiovaarae* and *Kluyveromyces marxianus* are also found in the *wadi* (Batra and Millner, 1974; Sandhu et al., 1986). The development and prevalence of microflora is affected by the seasons, summer being more favourable for bacteria and winter for yeasts.

The production of acid and gas results in a fall of pH from 5.6 to 3.2 and two-fold raise in the volume of the dough. The lactic acid bacteria are mainly responsible for the acidification of dough create favourable conditions for the yeasts to grow and become active for leavening. The fermentation brings about a significant increase in soluble solids, non-protein nitrogen, soluble nitrogen, free amino acids, proteolytic activity and vitamin B including thiamine, riboflavin and cyanocobalamine. On the other hand, the levels of reducing sugar and soluble protein decreases. Amylase activity increases initially, but declines thereafter (Sandhu and Soni 1989; Aidoo et al., 2005).

Papad

It is a pulse based condiment or savoury food and wafer like product originated in South India. The usual ingredients are Black gram, Bengal gram, lentil (*Lens culinaris*) and red gram or green gram, small quantity of peanut oil and common salt (Shupalekar, 1986). Black gram flour or a blend of black gram with Bengal gram, lentil (*Lens culinaris*), red gram or green gram (*Vigna radiata*) flour is hand kneaded with a small quantity of peanut oil, common salt (about 8%) and then pounded into a stiff paste. The dough is left to ferment for 1-6 hours. The fermented dough is shaped into small balls which are rolled into thin, circular flat sheets (10-24cm diameter, 0.2-1.2mm thick) and generally dried in the shade to 12-17% (w/w) moisture content. It is used to prepare curry or is eaten by itself as a crackly snack or appetizer with meals after roasting or deep frying in oil. The microorganism(s) involved are *Candida krusei* and *S. cerevisiae* (Aidoo et al., 2005; Roy et al., 2007).

Masyaura

It is a product made of Blackgram or greengram, *Colocosia* tuber, ashgourd or radish originated in Darjeeling hills and Sikkim. Blackgram is split in a traditional stone grinder (*Janto*) used for splitting and grinding grains. Cleaned, split blackgram is soaked in water overnight and the husk is removed by hand washing 2 to 3 times in water. The wet, dehulled blackgram is then ground in another traditional stone grinder (*Silauto*), which functions similar to the *Janto*. A small quantity of water is often added to aid grinding to get a thick paste. The *Colocosia* tuber is washed, peeled and finally shredded. The shreds are then mixed with blackgram dhal paste in approximate ratio of 1:1. The dough is then made into small lumps and spread on a tray. Usually, this is done in the evening and the trays are then left overnight to ferment at room temperature. The spongy, textured balls are then sun dried for 3 to 5 days and stored for use. The total microbial flora is dominated by lactic acid bacteria (>90%), *Pediococcus pentosaceous*, *Pediococcus acidilactic*,

Saccharomyces cerevisiae, Candida versatilis, Cladosporium sp., *Penicillium* sp. and *Aspergillus niger*. Dried *Masyaura* contains a final moisture content of about 8–10%, and is a rich source of protein (18–20%), carbohydrate (67–70%) and minerals (Karki, 1986; Dahal et al., 2005; Dahal et al., 2003).

Korean fermented soybean products

The most common Korean fermented soybean foods are chungkukjang, doenjang, kanjang (soy sauce), and kochujang. Chungkukjang is a short-term fermented soybean product similar to Japanese *natto*, whereas doenjang, kochujang, and kanjang (soy sauce) undergo long term fermentation as do Chinese tauchu and Japanese *miso*. Traditionally chungkukjang is made by fermenting the cooked but non-crushed soybeans for two or three days, it is usually prepared in autumn and winter after harvesting soybeans. Doenjang and kanjang are prepared by three step fermentations i.e first preparing the meju (dried soybean block) from cooked beans in late of October or early of November and ferment it for 1 or 2 months as solid fermentation; second aging meju for another 1 or 2 months in large earthenware jars by adding a salt solution as liquid fermentation, and then decanting the supernatant liquid to prepare kanjang from liquid and doenjang from the remaining soy paste; finally, both liquid and paste are aged for longer periods.

Kochujang is a unique and representative Korean traditional food for more than a thousand years (Kwon et al., 2011). Kochujang is usually prepared by mixing powdered red peppers, powdered meju, salt, malt-digested rice syrup, and rice flour, and the mixture is fermented for more than 6 months. Meju fermentation is the most important step in producing doenjang and kanjang, because many metabolite changes occur during this fermentation, whereas the changes in flavour, colour and bioactive components primarily depend on aging after fermentation.

Chungkukjang, doenjang, kanjang and kochujang are fermented with varying microorganisms when traditionally made, because fermentation conditions and ambient microbiota are different among regional environments. Meju can be fermented by traditional methods, being typically fermented by *Bacillus* species during the early stages of fermentation, followed by *Aspergillus oryzae*, which predominate during the remaining fermentation period. In traditionally fermented chungkukjang, many different bacterial strains were identified by rRNA gene sequencing or recA sequence using randomly amplified polymeric DNA-PCR method (Kwon et al., 2009), *Bacillus subtilis*, *B. amyloliquenfaciens*, *B. licheniformis* and *B. thuringiensis* were representative microorganisms during the chungkukjang fermentation. Among them, *B. subtilis* is the predominant strain in chungkukjang fermentation. It is well known that poly-α-glutamate (PGA) and nattokinase (fibrinolytic enzyme) are major components for maintaining a healthy immune system (Messina, 1995) and cardiovascular system (Sumi et al., 1987), respectively. Fibrinolytic microorganisms screened from chungkukjang, doenjang, meju and *natto*, were identified as *B. subtilis*, *Staphylococcussciuri*, *Enterococcus faecalis*, and *Citrobacter* or *Enterobacter* (Yoon et al., 2002).

Conclusion

Despite of the conventional foods largely produced from pulses in the Western world, there is a wide variety of products produced worldwide that have not received the scientific attention they deserve. These products are often fermented, and have an improved shelf-life and nutritional properties in comparison with the raw materials used. The flora responsible for the fermentation is in many cases is indigenous and includes strains of lactic acid bacteria, yeast and fungi. Single or mixed pulses sometimes mixed with other cereals are used, and the final texture of the product can vary according to the processing and fermentation conditions. Similar fermentation procedures have been used nowadays to develop new foods with enhanced health properties, which is a trend likely to continue in the future. A strategic research is needed to determine the relationship between fermentation and reduction in toxic compounds and to understand the dynamics of fermentation and how they can be manipulated for improvement of nutritional level and inhibition of food borne pathogens. It is imperative that modern biotechnological techniques be integrated into traditional processing procedures to upgrade their production methods.

Table 1: Traditional pulse based fermented foods consumed in various countries.

Sl.No.	Name of the product	Raw material used	Place of origin/use	Microbes associated	References
1	*Idli*	Rice and dehulled black gram	South India	*Leuconostoc mesenteroides, Streptococcus faecalis, Lactobacillus delbrueckii, Lactobacillus fermenti, Lactobacillus lactis Pediococcus cerevisiae*	Nagaraju & Manohar, 2000 Purushothaman, et al.,1993; Ramakrishnan, 1993
2	*Dosa*	Rice and dehulled black gram	South India	*Leuconostoc mesenteroides, Streptococcus faecalis, Torulopsis candida, T. pullulans*	Battacharya & Bhat, 1997 Chavan & Kadam, 1989; Purushothaman et al., 1993
3	Soy sauce	Soybeans and wheat	Japan, China, Taiwan	*Aspergillus oryzae and A.soyae*	Beuchat, 1983; Xu, 1990; Yokotsuka & Sasaki, 1998 Allen et al., 1996; Hamada, et al., 1991, Holzapfel, et al., 1998; Roling, et al., 1996
4	*Tempeh*	Soybeans	Indonesia	*Rhizopus oligosporus*	Winaro and Reddy 1986
5	*Miso*	Rice and soy beans	Japan, China	*Aspergillus oryzae, Torulopsis etchellsii, Lactobacillus*	Ebine 1986
6	*Natto*	Soybean	Japan	*Bacillus subtilis natto*	Sumi et al., 1987 Sumi, 1998
7	*Kinema*	Yellow soybean seeds	Darjeeling, Sikkim	*Bacillus subtilis, B. licheniformis, B. circulans, B. thuringiensis, B. sphaericus, Enterococcus faccium, Candida parasilosis Geotrichum candidum*	Sarkar et al., 1993, Sarkar and Tamang 1994, Sarkar et al., 1996, Tamang and Nikkuni 1998

Sl.No.	Name of the product	Raw material used	Place of origin/use	Microbes associated	References
8	*Tungrymbai*	Soybean	Meghalaya	*Bacillus subtilis, LAB and yeast*	Dike and Odunfa, 2003, Sohliya et al., 2009
9	*Axone or Akhone* or *Kheuha or Sabrocha* or *Sadocha.*	Soybean	Northern east India.	*Bacillus subtilis*	Singh et al., 2007(b), Mao and Odyuo, 2007; Tamang et al., 2009
10	*Bekang*	Soybean	Mizoram	*Bacillus subtilis*	Singh et al., 2007, Tamang et al., 2009
11	*Aagya*	Soybean	Arunachal Pradesh	--	Singh et al., 2007
12	*Peron naming*	Soybean	Arunachal Pradesh	--	Singh et al., 2007
13	*Yanni perung*	Soybean	Arunachal Pradesh	--	Singh et al., 2007
14	*Hawaizaar*	Soybean	Manipur	--	Singh et al., 2007
15	*Peruyyan*	Soybean	Arunachal Pradesh	*Bacillus subtilis, Bacillus sp. and Lactic Acid Bacteria*	Tamang et al., 2009
16	*Soybean (libi) Chhurpi*	Soybean	Arunachal Pradesh	--	Singh et al., 2007
17	Wadi	Black gram	Punjab and West Bengal	*L. mesenteroides, L. fermentum, S. cerevisiae, Trichosporon cutaneum, Candida vartiovaarae, Kluyveromyces marxianus*	Batra and Millner 1974, Sandhu et al., 1986. Sandhu and Soni 1989, Aidoo et al., 2005
18	Papad	Black gram, Bengal gram, lentil (*Lens culinaris*) red gram or green gram	South India	*Candida krusei and S. cerevisiae*	Shupalekar, 1986, Aidoo et al., 2005 Roy et al., 2007

Sl.No.	Name of the product	Raw material used	Place of origin/use	Microbes associated	References
19	*Masyaura*	Blackgram or greengram	Darjeeling and Sikkim	*Pediococcus pentosaceous, Pediococcus acidilactic, and Lactobacillussp. Saccharomyces cerevisiae and Candida versatilis, Cladosporium sp., Penicillium sp. and Aspergillus niger.*	Karki, 1986, Dahal et al., 2005, Dahal et al., 2003
20	*Channa dal dhokla*	*Channa dal*	North India	*Leuconostoc mesenteroides, Streptococcus faecalis Torulopsis candida, T. pullulans*	Chavan & Kadam, 1989; Purushothaman et al., 1993 Aliya & Geervani, 1981;
21	Chungkukjang, doenjang, kanjang and kochujang.	Soybean	Korea	*Aspergillus oryzae Bacillus subtilis, B. amyloliquenfaciens, B. licheniformis and B. thuringiensis*	

References

Adams MR (1990) Topical aspects of fermented foods. Trends Food Sci Technol 1:141-144.

Agrahar-Murungkar D and Subbulakshmi G (2006) Preparation techniques and nutritive value of fermented foods from the Khasi tribes of Meghalaya. Ecol Food Nutr 45: 27–38

Agropedia (2016) http://agropedia.iitk.ac.in/content/area-production-and productivity-major-pulses. Accessed 1 May 2016.

Aidoo KE, Nout MJR and Sarkar PK (2006) Occurrence and function of yeasts in Asian indigenous fermented foods. *FEMS Yeast Res* 6(1):30–39.

Alexander JW, Boyce ST and Babcock GF (1990) The process of microbial translocation. Ann Surg 21(2):496-512.

Aliya S and Geervani P (1981) An assessment of protein quality and vitamin B content of commonly used fermented products legumes and millet. J Sci Food Agric 32:837–842.

Allen WD, Linggood MA and Porter P (1996) Probiotic. US Pat.No. 5,589,168 1-12.

Ayres R, Steinkraus KH, Olek A and Farr D (1993) Biochemistry of saccharomyces. In: Steinkraus K H (ed) Handbook of indigenous fermented foods, Marcel Dekker, New York, p 517–519.

Batra LR and Millner PD (1974) Some Asian fermented foods and beverages and associated fungi. Mycologia 66: 942 – 950.

Battacharya S and Bhat KK (1997) Steady shear rheology of rice blackgram suspensions and suitability of rheological models. J Food Engg 32:241–250.

Beuchat LR (1983) Indigenous fermented foods. In: Reed G (ed) Biotechnology: Food and feed production with microorganisms, Verlag Chemie, Weinheim.

Billings T (1998) On fermented foods. Available: http://www.livingfoods.com.

Carrico J and Meakin JI (1986) Multiple organ failure syndrome. Arch Surg 121:196-208.

Case S and Fenster C (2011) Pulses and the gluten free diet. Pulse Canada, Canada, p 6-9.

Chavan JK and Kadam SS (1989) Critical reviews in food science and nutrition. Food Sci 28:348–400.

Dike EN and Odunfa SA (2003) Microbiological and biochemical evaluation of a fermented soyabean product soya dawadawa. J Food Sci Technol 40:606-610.

dpd.gov.in (2022) Annual Report 2021-22. https://dpd.gov.in/Annual%20Report%20(2021-22).pdf. Accessed on 2/4/23

Duszkiewicz-Reinhard W, Gujska E and Khan K (1994) Reduction of stachyose in legume flours by lactic acid bacteria. J Food Sci 59:115-117.

Franta B, Steinkraus KH, Mattick LR, Olek A and Farr D (1993) Biochemistry of saccharomyces. In: Steinkraus KH (ed) Handbook of indigenous fermented foods, Marcel Dekker, New York, p 517–519.

Fuller R (1989) Probiotica in man and animals. J Appl Bacteriol 66.

Goldin BR and Gorbach SL (1984) The effect of oral administration on *Lactobacillus* and antibiotics on intestinal bacterial activity and chemical induction of large bowel tumors. Dev Ind Microbiol 25:139-150.

Gopalan C, Rama Sastri BV and Balasubramanian SC (2004) Nutritive Value of Indian Foods, National Institute of Nutrition, ICMR, Hyderabad.

Guermani L, Villaume C, Bau HW, Chandrasiri V, Nicolas JP and Mejean L (1992) Composition and nutritional value of okara fermented by Rhizopus oligosporus. Sciences des Aliments 12:441-451.

Gupta V and Nagar R (2010) Effect of cooking, fermentation, dehulling and utensils on antioxidants present in pearl millet rabadi – a traditional fermented food. J Food Sci and Technol 47(1):73-76.

Gupta V, Sharma A and Nagar R (2007) Preparation, acceptability and nutritive value of rabadi – a fermented moth bean food. J Food Sci and Technol 44(6): 600-601.

Gustafsson BE (1983) Introduction to the ecology of the intestinal microflora and its general characteristics. In: Hallgren B (ed) Nutrition and the Intestinal Flora, ISBN 91 22 00593 5.

Hamada T, Sugishita M, Fukushima Y, Fukase T and Motai H (1991) Continuous production of soy sauce by a bioreactor system. *Process Biochem*, 26:39–45.

Harlander S (1992) Food biotechnology. In: Lederberg J (ed) Encyclopedia of microbiology, Academic Press, New York, p 191–207.

Hesseltine CW (1983) The future of fermented foods. Nutr Rev 41: 293-301.

Holzapfel WH, Haberer P, Snel J, Schillinger U and Veld JH (1998). Overview of gut flora and probiotics. Int J Food Microbiol 41:85–101.

Horitsu H, Wang MY and Kawai K (1991) A modified process for soy sauce fermentation by immobilized yeasts. Agri Biol Chem 55:269–271.

Hosono A and Hisamatsu S (1995) Binding of amino acid pyrolysates and aflatoxins to autoclaved cells of *Enterococcus faecalis* FK-23. Biosci Biotechnol Biochem 59:940-942.

Jeyaram K, Singh A, Romi W, Devi AR, Singh WM, Dayanithi H, Singh NR and Tamang JP (2009) Traditional fermented foods of Manipur. Ind J Trad Knowledge 8(1): 115-121.

Kasravi FB, Adawi D, Molin G, Bengmark S and Jeppson B (1997) Effect of oral supplementation of lactobacilli on bacteria translocation in acute liver injury induced by D-galactosamine. J Hepatol 26:417-424.

Khetarpaul N and Chauhan BM (1990) Fermentation of pearl millet flour with yeasts and lactobacilli: in vitro digestibility and utilisation of fermented flour for weaning mixtures. Plant Foods Hum Nutr 40:167-173.

Khetarpaul, N and Chauhan BM (1991) Biological utilization of pearl millet flour fermented with yeasts and lactobacilli. Plant Foods Hum Nutr 41:309-319.

Kiers JN, Nout MJ and Rombouts FM (2000) In vitro digestibility of processed fermented soybean, cowpea and maize. J Sci Food Agric 80:1325-1331.

Mao AA and Odyuo N (2007) Traditional fermented foods of the Naga tribes of Northeastern India. Ind J Trad Knowledge 6(1): 37-41.

McIntosh GH (1996) Probiotics and colon cancer prevention. Asia Pac J Clin Nutr 5:48-52.

McKay LL and Baldwin KA (1990) Applications for biotechnology: present and future improvements in lactic acid bacteria. FEMS Microbiol Rev, 87:3–14.

Mensah P (1997) Fermentation—the key to food safety assurance in Africa? Food Control 8:271–278.

Ministry of Agriculture and Farmers Welfare (2016) http://dacnet.nic.in/eands/latest_2006.htm. Accessed 1 May 2016.

Mital BK and Garg SK (1995) Anticarcinogenic, hypoholesterolemic and antagonistic activities of *Lactobacillus acidophilus*. CRC Crit Rev Microbiol 21:175-214.

Murthy CT and Rao PNS (1997) Thermal diffusivity of Idli batter. J Food Engg 33:299–304.

Murthy CT, Nagaraju VD, Rao PNS and Subba Rao VN (1994) A device for continuous production of Idli. Indian patent no.950/DEL/94.

Nagaraju VD and Manohar B (2000) Rheology and particle size changes during Idli fermentation. J Food Engg 43:167–171.

Nout MJR (1995) Useful role of fungi in food processing. In : Samson RA, Hoekstra E, Frisvad JC and Filtenborg O (eds) Introduction to food-borne fungi, Central Bureau voor Schimmel cultures, Baarn, p 295–303.

Nout MJR and Sarkar PK (1999) Lactic acid food fermentation in tropical climates. Antonie van Leeuwenhoek, 76: 395-401.

Nout MJR, Bakshi D and Sarkar PK (1998) Microbiological safety of kinema, a fermented soya bean food. Food Control 9(6): 357–362.

Obizoba IC and Atii JV (1991) Effect of soaking, sprouting, fermentation and cooking on nutrient composition and some anti-nutritional factors of sorghum (Guinesia) seeds. Plant Foods Hum Nutr 41:203-212.

Paredes-López O and Harry GI (1988) Food biotechnology review: traditional solid-state fermentations of plant raw materials - application, nutritional significance and future prospects. CRC Crit Rev Food Sci and Nutr 27:159-187.

Pulses and Legumes - The Nutrient Rich Seed (2016) http://food.sify.com/articles/Pulses_And_Legumes_The_Nutrient_Rich_Seed-182722. Accessed 15 May 2016.

Purushothaman D, Dhanapal N and Rangaswami G (1993) Indian idli, dosa, dhokla, khaman, and related fermentations. In :Steinkraus KH (ed) Handbook of indigenous fermented foods, Marcel Dekker, New York, p 149–165.

Ramakrishnan CV (1993) Indian idli, dosa, dhokla, khaman, and related fermentations. In: Steinkraus KH (ed) Handbook of indigenous fermented foods, Marcel Dekker, New York, p 149–165.

Reddy BS, Ekelund G, Bohe M, Engle A and Domellöf L (1983) Metabolic epidemiology of colon cancer: Dietary pattern and fecal sterol concentrations of three populations. Nutr Cancer 5, 34-40.

Robinson RK and Samona A (1992) Health aspects of 'bifidus' products: a review. Int J Food Sci Nutr 43:175-180.

Roling WFM, Apriyantono A and Van Verseveld HW (1996) Comparison between traditional and industrial soy sauce (Kecap) fermentation in Indonesia. J Ferm Bioengg 81:275–278.

Rowan NJ, Anderson JG and Smith JE (1998) Potential infective and toxic microbiological hazards associated with the consumption of fermented foods. In: Wood JB (ed) Microbiology of fermented foods, Blackie Academic and Professional, London, p 812–837.

Roy A, Moktan B and Sarkar PK (2007) Microbiological quality of legume-based traditional fermented foods marketed in West Bengal, India. Food Control 18:1405–1411.

Salminen S, Deighton M and Gorbach S (1993) Lactic acid bacteria in health and disease. In: Salminen S and Wright A von (ed) Lactic acid bacteria, Marcel Dekker Inc., New York, USA, p 237-294.

Sandhu DK and Soni SK (1989) Microflora associated with Indian Punjabi warri fermentation. J of Food Sci Technol 26: 21-25.

Sandhu DK, Soni SK and Vilkhu KS (1986) Distribution and role of yeasts in Indian fermented foods. In: Vashishat RK and Tauro P (eds) Yeast Biotechnology, Haryana Agricultural University, Hisar, India, p 142 – 148.

Sands DC and Hankin L (1974) Selecting lysine-excreting mutants of lactobacilli for use in food and feed enrichment. J ApplMicrobiol 28, 523–534.

Sarkar PK, Cook PE and Owens JD (1993) *Bacillus* fermentation of soybeans. World J Microbiol Biotechnol 9(3): 295-299.

Sarkar PK, Jones LJ, Craven GS, Somerset SM and Palmer C (1997) Amino acid profiles of kinema, a soybean fermented food. Food Chem 59(1): 69-75.

Sarkar PK, Tamang JP, Cook PE and Owens JD (1994) Kinema – a traditional soybean fermented food: Proximate composition and microflora. Food Microbiol 11: 47-55.

Sarkar PK, Hasenack B and Nout MJR (2002) Diversity and functionality of Bacillus and related genera isolated from spontaneously fermented soybeans (Indian Kinema) and locust beans (African Soumbala). *Int J Food Microbiol* 77(3):175-186.

Sarkar PK and Tamang JP (1994) The influence of process variables and inoculum composition on the sensory quality of kinema. Food Microbiol 11(4):317–325.

Savage DC (1983) Microbial ecology of the gastrointestinal tract. In: B. Hallgren B (ed) Nutrition and the Intestinal Flora, ISBN 91 22 00593 5.

Scheinbach S (1998) Probiotics: functionality and commercial status. Biotechnol Advances, 16, 581–608.

Sekar S and Mariappan S (2007) Usage of traditional fermented products by Indian rural folks and IPR. Ind J Trad Know, 6 (1): 111-120.

Shahani KM (1983) Nutritional impact of lactobacillic fermented foods. In: Hallgren B (ed) Nutrition and the Intestinal Flora, ISBN 91 22 00593 5.

Sharma A and Kapoor AC (1996) Level of antinutritional factors in pearl millet as affected by processing treatments and various types of fermentation. Plant Foods Hum Nutr 49: 241-252.

Shortt C (1998) Living it up for dinner. Chem Ind, 8,300–303.

Shurpalekar SR (1986) Papads. In: Reddy NR, Pierson MD and Salunkhe DK (eds) Legume-based Fermented Foods, CRC Press, Boca Raton, FL, p. 191 – 217.

Simango C (1997) Potential use of traditional fermented foods for weaning in Zimbabwe. J Social Sci Med 44:1065–1068.

Singh A, Singh RK and Sureja AK (2007) Cultural significance and diversities of ethnic foods of Northeast India. Ind J Trad Know 6(1):79-94.

Sohliya I, Joshi SR, Bhagobaty RK and Kumar R (2009) Tungrymbai – A traditional fermented soybean food of the ethnic tribes of Meghalaya. Ind J Trad Know 8(4): 559 -561.

Soni SK and Sandhu DK (1990) Indian fermented foods: microbiological and biochemical aspects. Ind J Microbiol 30: 135–157.

Stanbury PF (1999) Fermentation Technology. In: Stanbury PF, Whitaker A, and Hal SJ (eds) Principles of Fermentation Technology, 2nd Edition, Butterworth Heinemann, Oxford, UK, p 1-24.

Steinkraus KH (1998) Bio-enrichment: production of vitamins in fermented foods. In Wood JB (ed.) Microbiology of fermented foods, Blackie Academic and Professional, London, p. 603–619.

Steinkraus KH, Ayres R, Olek A and Farr D (1993) Biochemistry of Saccharomyces. In: Steinkraus KH (ed) Handbook of indigenous fermented foods, Marcel Dekker, New York, p 517–519.

Sumi H (1994) Structure and fibrinolytic properties of nattokinase. In: Sumi H (ed) Basic and clinical aspects of Japanese traditional food Natto, Japan Technology Transfer Association, p 49-56.

Sumi H (1998) Accumulation of Vitamin K (Menaquinone-7) in Plasma after Ingestion of Natto and Natto Bacilli (*B. subtilis natto*). Kurashiki, Okayama Japan: Japan Functional Food Research Association, Department of Physiological Chemistry, Kurashiki University of Science and the Arts, p. 1–5.

Sumi H, Hamada H, Tsushima H, Mihara H and Muraki H (1987) A novel fibrinolytic enzyme (nattokinase) in the vegetable cheese natto: A typical and popular soybean food in the Japanese diet. Experientia 43(10):1110–111.

Tamang JP (2003) Native microorganisms in the fermentation of kinema. Ind J Microbiol 43 (2): 1-4.

Tamang JP and Nikkuni S (1998) Effect of temperatures during pure culture fermentation of kinema. World J Microbiol Biotechnol 14: 847-850.

Tamang JP, Chettri R and Sharma RM (2009) Indigenous knowledge on North-East women on production of ethnic fermented soybean foods. Ind J Trad Know, 8 (1):122-126.

Tang D, Dong Y, Ren H and He C (2014) A review of phytochemistry, metabolite changes, and medicinal uses of the common food mung bean and its sprouts (*Vigna radiata*). Chem Cent J 8:4

Teniola O D and Odunfa SA (2001) The effects of processing methods on the levels of lysine, methionine and the general acceptability of ogi processed using starter cultures. Int J Food Microbiol 63:1–9.

Urooj A and Puttaraj S (1994) Effect of processing on starch digestibility in some legumes – An in vitro study. Nahrung 38:38-46.

Wang HL and Fang SF (1986) History of Chinese fermented foods. In: Hesseltine CW and Wang HL (eds) Indigenous fermented food of non-western origin, Cramer, Berlin, p 23–35.

Wells CL (1990) Relationships between intestinal microecology and the translocation of intestinal bacteria. Antoine van Leeuwenhoek 58:87-93.

Winaro FG and Reddy NR (1986) Legume-Based Fermented Foods. CRC Press, Inc., Boca Raton, FL.

www.ag.ndsu.edu (2012) Developed for the Northern Pulse Growers Association by Julie Garden-Robinson. Accessed 1 May 2013

Xu Y (1990) Advances in the soy sauce industry in China. J Ferm Bioengg 70:434–439.

Yeoh QL (1995) A fermentation process for the improvement of soya sauce production. Asian Pac J Mol Biol Biotechnol 3:311–316.

Yokotsuka T (1993) Japanese shoyu. In: Steinkraus KH (ed.) Handbook of indigenous fermented foods, Marcel Dekker, New York, p. 511–517.

Yokotsuka T and Sasaki M (1998) Fermented protein foods in the orient: shoyu and miso in Japan. In: Wood J B (ed.) Microbiology of fermented foods, Blackie Academic and Professional, London, pp. 351–415.

Yong FM and Wood BJB (1974) Microbiology and biochemistry of soy sauce fermentation. Advances Appl Microbiol 17:157–230.

Yong FM and Wood BJB (1976) Microbial succession in experimental soy sauce fermentations. J Food Technol 11:525–536.

Yong FM and Wood BJB (1977) Biochemical changes in experimental soy sauce koji. J Food Technol 12:163–175.

12

Role of Bioinformatics in Food Industry

Aishwarya Vetrivel and Rajeswari Murugesan

Department of Biochemistry, Biotechnology and Bioinformatics
Avinashilingam Institute for Home Science and Higher Education for Women
Coimbatore, Tamil Nadu

Abstract

Bioinformatics is an interdisciplinary field which plays a major role in various disciplines including drug discovery, pharmacogenomics, proteomics, comparative genomics, microbial genome applications in biotechnology. Food plays a vital role in various processes in body and also helps in controlling various chronic diseases. Food safety and food processing are very crucial factor for human lives. Bioinformatics has been successfully applied in food industry for improving various factors of food such as flavor, taste, allergenicity, probiotics, optimizing the quantitative and compositional parameters for food processing and assessing the food safety. In this chapter, we have given an overview about the role of bioinformatics in food industry.

Introduction

In present scenario, the demand from consumers is increasing day by day in the want of healthy, safe, and nutritious food that has minimal synthetic preservatives in it. According to the reports of World Health Organization (WHO, 2019), around 600 million people are affected by foodborne illness globally every year, with approximately 420,000 deaths (Hoffmann et al., 2019). In such a case, ensuring food quality and safety is an area of concern for food scientists, food industries, and regulatory authorities worldwide. The quality of food product is significantly deteriorated as the food commodities travels through various stages that make them prone to attack of various extrinsic and intrinsic factors (Prakash et al., 2020). The existing methods of identification of biological contaminants, and preservation, and their related risks have its own limitations, are expensive, and insufficient.

In this context, emerging advances in omics science could be utilized for investigation, detection, control, and prevention of foodborne toxins, and microbes. Hence the foodomics which is a holistic integrated approach of several omics technologies referred to as omics in food science has gained a significant popularity by the regulatory authorities, and industries to prevent the issues related to food safety, traceability, bioactivity, and quality. Foodomics approaches includes genomics, transcriptomics, proteomics, and metabolomics that explore the mechanism of RNA, gene expression, metabolites, and proteins related to food at molecular level (Cifuentes, 2017).

Over two decades ago, the word "bioinformatics" combined experimental and computational technology to make a significant contribution to scientific research. Bioinformatics has thought to bring a standard shift among various scientific discipline namely comparative genomics, microbial genome applications, molecular evolution, molecular medicine, biotechnology, and drug discovery. Also, the research concerned to this science field predominantly handle various thrust areas namely sequence analysis, modeling biological networks, genome sequencing, annotation, comparison, protein structure prediction, protein allergenicity detection, and expression data analysis. (Xu, 2018; Holton et al., 2013). Recently, many bioinformatic technologies have given their support venture to protect public health. In particular, various computer technologies and emerging science has led to the improvisation of safety in food production and food processing through innovative methods (Healy et al., 2016).

Out of necessity, bioinformatics has become a growing area in science and food research. Combining the tools of information technology, genomic data, and other advanced techniques could help biologists to investigate about all the components of a biological system at the same time efficiently. Relatively food science is also a systematic and coherent technology which could integrate knowledge of various sciences such as chemical, physical, biological, and biochemical properties of food materials (Rani and Gomathi, 2017).The bioinformatics approaches related to its application in food sciences has become a significant tool for research in life sciences as it integrates large amount of data obtained from comparative genomics and proteomics in food research to traditional techniques (Waidha et al., 2016). In addition, bioinformatics tools can provide interpretation of data obtained from "foodomics" associated with flavour, taste, food quality, and safety (Holton et al., 2013).

The emergence of scientific advancements in the field of genome research and exploitation of biotechnological techniques could represent significant contribution to increase food performance and provide deeper information about their multiple functionalities. Prior to bioinformatics and genomics era, biological effects was investigated by designing suitable markers for a particular condition. In this new era, with the increase in biological knowledge, consistent improvements in health related to certain metabolic conditions will be the new markers. Hence, this insight understanding may intervene through foods to control health problems prior to its effects that may be apparent and lead consumers to get an advantage in areas of technological breakthrough which would result in high quality, healthy foods with positive nutritive properties (Rani and Gomathi, 2017).

Several issues related to food safety like preparation, handling, and storage of foods should be taken into consideration to control foodborne illness and to prevent associated health hazards. However, it is still ambiguous on how the tools of bioinformatics aid in the safety evaluation and food industry for achieving high quality and production (Xu, 2018). With this background, this chapter will primarily focus on the advances, applications, role, and future possibilities of bioinformatics in food industry.

Omics Approaches in Food Science

In recent years, there is an invaluable advancement in the research based on food safety science through the bioinformatics and genomics approaches. Although there is a massive emergence, their applications has been limited in the area of research development on molecular diagnostic techniques. For instance, bioinformatics and genomics plays a significant role in development of standard molecular technologies that are routinely used for identification and tracking of foodborne diseases in laboratories. The molecular technologies that are of concern in this case includes Multi-Locus Variable-Number of Tandem Repeats Analysis (MLVA), Pulse-Field Gel Electrophoresis (PFGE), and Multi-Locus Sequence Typing (MLST). These tests needs only substantial bioinformatics and genomics for development in such a way, they are the modest genomics and bioinformatics technologies to be carried out. These tests have been historically incorporated with genomics and bioinformatics into various software systems and standard lab protocols, effectively hidden out of mind and sight to wide range of users (Taboada et al., 2017).

The technology aims to decipher the biological process mechanistically by innovation of novel methods and software as tools for processing, analysing, and interpretation of biological data from various omics and biochemical approaches. According to existing literature, there is a positive correlation between functional food ingredients and disease conditions. Thus, considerable interest in foodomics technology has gained interest worldwide in the field of nutrition and food science (Valdés et al., 2017).

Food safety has become one of the major concern for food industries and consumers and have established a unique research based program to enhance food safety with well-standardized procedures as a state of new research tools. The presently available genomics era with the aid of advanced tools like proteomics and DNA-array technology could be able to solve the problems related to food safety in an effective way. Hence, these new technologies have become a crucial part of the modern scientific plan to provide assurance in food safety to consumers (Rani and Gomathi, 2017).

In the field of food science, bioinformatics occupies a predominant position in prediction and assessment of both desirable and undesirable effects of microbes on food, proteomics and genomics analysis to fulfil the necessity in food processing, food production, quality improvisation, enhancement of nutritive value of various food sources and so on. Additionally, these omics approaches can also be applied for production of good quality crops which can lead to disease resistant and high yield in crop varieties (Kumar and Chordia, 2017).

There is also an increase in need for rapid identification technique for possible contaminants found in food and its negative effect on food ingredients, nutrients, and health of consumers in the past few years. Hence, the interest of the scientific community in the new discipline of science regarded as "foodomics" has been inevitably at its rise (Alvarez-Rivera et al., 2018). The major omics approaches includes genomics, transcriptomics, proteomics, and metabolomics which could provide adequate data and information to control the problems related to food safety rationally.

Genomics is an area that deals with the study of the gene, their expression, and function. It provides a wider information about the biological contaminants found in food and its shelved products. With the advent of modern, cutting-edge genomics based technologies like whole-genome sequencing (WGS), the cost and time for the study of microbial genome are drastically decreased. The next generation sequencing approach could be utilized for monitoring, inspection, identification, and assessment of foodborne pathogens based on a genetic blueprint with a preventive measure for future outbreaks (World Health Organization, 2019).

Transcriptomics deals with the study of complete set of RNA transcripts produced by the genome referred to as transcriptome. Comparing transcriptomes could decipher the gene expression responsible for synthesizing biologically active compounds, resistance/tolerance against the adverse environmental conditions, toxin production, responses of microbes to physical, chemical, or biological stress, nutrients, and biofilm formation (Mehta and Hasija, 2018).

On the other hand, proteomics deals with comparison of expression of proteomes both qualitatively and quantitatively at various environmental conditions at a particular time. In food science, it provides a detailed account about structure, expression, function, and interaction of protein with other biologically active components in the food matrix (Bendexin et al., 2011). This technique could be used as a biomarker for identification of foodborne pathogenic microbes based on its metabolite's products and specific proteins (Prakash et al., 2020).

Among various omics approaches, metabolomics analysis posed a significant impact on food safety as the quality and acceptability of food products are greatly affected by the extrinsic and intrinsic environmental conditions which are influenced by the metabolite production of organism. Thus, nowadays food scientists focus on the use of metabolomic approaches to characterize toxic metabolite products found in food (Valdés et al., 2017).

In this context, Global Coalition for Regulatory Science Research (GCRSR) was established in 2013, with various international collaboration stepped forward to investigate data from different science-based technologies and translated into regulatory application (Tong et al., 2015). Molecular docking studies and structural bioinformatics has revolutionized this area by investigating the biological response of chemicals that activates corresponding receptors and elicit specific response (Holton et al., 2013). Fang et al. (2011) reported various tools of bioinformatics could be used in conjunction with next generation sequencing technologies for rapid identification and investigation of foodborne pathogens.

Nowadays, to understand the functional aspects of biologically active compounds a wide range of bioinformatics tools are exploited including protein databases (NCBI (https://www.ncbi.nlm.nih.gov/) UniProtKB (https://www.uniprot.org/), BIOPEP (http://www.uwm.edu.pl/biochemia/index.php/pl/biopep)) for sequence and structural information of various proteins; allergenicity/toxicity prediction/analysing tools including AlgPre (http://crdd.osdd.net/raghava/algpred/), ToxinPred (http://crdd.osdd.net/raghava/toxinpred/), BIOPEP to predict proteins associated with toxicity and allergies of proteins; Protein Structure Prediction Server (I-TASSER) (https://zhanggroup.org/I-TASSER/) to study protein's structure and its bioactivity prediction by AHTPDB (http://crdd.osdd.net/raghava/ahtpdb/), BIOPEP, Peptide Ranker (http://distilldeep.ucd.ie/PeptideRanker/), AntiBP2 (http://crdd.osdd.net/raghava/antibp2/) which could pave way for the identification of antibacterial peptides, antihypertensive peptides, and precursor proteins. Other *in silico* digestion tools such as Peptide Cutter (https://web.expasy.org/peptide_cutter/), BIOPEP, Enzyme Predictor (http://bioware.ucd.ie/~enzpred/Enzpred.php) may serve helpful for elucidating the potent cleavage sites in protein sequence (Tu et al., 2018).

Another research tool that has revolutionized over the last decade is the next generation sequencing (NGS) applied routinely in various fields for investigations, outbreak, forensics, diagnostics, and food authenticity (Allard et al., 2017). This technology is rapidly at rise and have greater influence on food microbiology because of its cost reduction and quality enhancement. NGS works in food microbiology in two different ways either as a tool to determine the entire genome of a single cultured isolate referred to as whole genome sequencing or applied to biological samples for obtaining sequences from multiple organisms referred to as metagenomics. This technique has gained acceptance as surveillance tool for assessment of foodborne pathogens because of its efficient discriminatory power compared to traditional molecular typing tools (Jackson et al., 2016).

Although there has been a huge advancements for the past years, the speed in providing retrieved information to industries has drastically limited the adoption of WGS technology for the purpose of foodborne pathogens. The additional value of WGS technology over traditional methods of molecular typing techniques depends on interpretation of the results accurately. WGS is considered a powerful tool for pathogen source tracking which determines the relatedness of bacterial isolates with a limitation in which isolates that arose from the same source could be indicated. Integration of this information obtained through WGS technology to that of its likely origin of contaminants in food, appropriate expertise, contextualization of information obtained from results of that sample known as metadata are the most important characteristics needed for a researcher in this field. Thus, WGS has the potential to provide information that will eventually be exploited by quality managers to generate data driven strategies for management of food safety (Barretto et al., 2021).

Comparative genomics, another extremely diverse field could analyse similarities and variations between genomes. These approaches provide information about the size and structure of genomes in the domain of life (Bentley & Parkhill, 2004), decipher the unique genetic elements (Khaldi and Wolfe, 2011; Ma et al., 2010), modes of evolution (Ogura et al., 2009), adaptations to different environments (Omelchenko et al., 2005), and components that are genetically involved in pathogenicity capacities (Chun et al., 2009; Khaldi et al., 2010).

Practically, this field of research has gained significance in assessing the regulatory genes responsible for production of biofuel in fungi (Wohlbach et al., 2011). Various food pathogens, food-based sources, are the primary focus for comparative genomic and sequencing projects (Mayer et al., 2011). These projects contributed to a better method for control of diseases related to crops (Qian et al., 2005), poisoning of food (Abee et al., 2004), and also augmented the knowledge of properties including ripening (Fei et al., 2004), flavour formation (Liu et al., 2008) and improvement of crops (Mochida & Shinozaki, 2010).

Greater attempts are now focused on the application of these technologies to supplement the nutritive efficacy and quality of various food sources. Hence, food science has revolutionized anonymously to embrace growing technologies of "omics" suite (Holton et al., 2013). Although there is a reasonable acquisition of bioinformatics in these fields, various issues stands outstanding that remains as conveyance towards computation continues.

Initially, there is a want for education to understand its potentiality and awareness of its limitations related to the application of bioinformatics in food and nutrition. Secondly, there should be an introduction to researchers with unique backgrounds on the common and unified language to communicate from. Finally, there should be some knowledge about combined database that can store and prune wide range of resources and biological data which are considered important to food sciences. These measures could be vital for present development of bioinformatics in food sciences and thereby provide an expedient and simple way for researchers to take up a new discipline (Holton et al., 2013; Lemay et al., 2007).

Food Composition Databases

Bioinformatics community to date has not explored extensively about the notable food composition databases that are in effect globally. But here we have tried to incorporate some of the databases that are available for use. These databases are tools which are very much essential for assessment of nutritional quality by health scientists, typically on a regional or national basis (Pennington et al., 2007). USDA National Nutrient Database for Standard Reference (NNDSR) (http://www.ars.usda.gov/ba/bhnrc/ndl) also known as foodomics database is a major food composition database found in United States, available as a free and open access resource. The database is updated on a regular basis and feature data's for around 8000 foods currently, which makes them as one of the most comprehensive food composition databases for assessment of nutrients globally (Merchant and Dehghan, 2006).

EuroFIR (http://www.eurofir.org/) is another standardized food composition database native to Europe beneficial to healthcare and food scientists. EuroFIR enlists software developers among their users to demonstrate the succession of research in this field based on computational techniques. This database is maintained by a non-profit organization and requires a membership to acquire data. Finally, the FAO/INFOODS Analytical Food Composition Database (http://www.fao.org/infoods/infoods/tables-and-databases/faoinfoods-databases/en/) issue access for food scientists to obtain information on composition of foods that are consumed commonly worldwide (Holton et al., 2013; Gry et al., 2007).

FooDB is another comprehensive resource database that is concerned with chemistry, biology, and food constituents. It contains information about the macronutrients and micronutrients whose constituents are essential for food to give taste, texture, colour, aroma, and flavour. It is accessible online as a free source. Each of the chemical entry in this database covers detailed information about the biochemical, compositional and physiological parameters of food. The data found in this database are obtained from the literature and includes chemical class of compounds, nomenclature, description, information on its structure and physicochemical characteristics, concentrations, food sources, taste, aroma, colour, health and physiological effects on various foods (Kumar and Chordia, 2017). This type of extensive research in this area serves as a unique tool to substantiate nutritional and food sciences. Further development of this research beyond data on composition of food could greatly lead to advancement in bioinformatics based technological research in these fields.

Although various databases are available publicly, the requirement for application of bioinformatics are to handle multiple data accurately and in a uniform manner. For this purpose, data has to be kept in a fixed and precise arrangement or configuration for convenient observation by computational biology. This type of arrangement and integration of nutritional and food data allows comparative analysis which is the most important tool for the discovery of new opportunities and directions. To support this formatting, FoodWikiDB (http://www.foodb.ca/) was considered as a storage platform for food based information to ease the use of computational tools. Additionally, a wiki-like food database can serve as an appropriate platform for bioinformaticians to explore comparative and integrative approaches between various data types, and to predict novel experimental outcomes (Holton et al., 2013).

Microbial Informatics

This area of research highlighted an enormous role in prediction and assessment of desirable and undesirable possessions of microbes on food products. Integration of data obtained from phenotypes, genomics, systems biology, and biomarkers with access for databases to predict sequence-related microbial functionality and for functionality pertained to specific taxa could investigate the effects of microbes on foods. Applying the technique of microbial bioinformatics, the yield of biomass can be well optimized which could thereby influence the overall production of food effectively.

For this purpose, several metabolic models and sequence information of the organism's genome have been developed for various microbes that majorly incorporated food-based microorganisms. The complete genome-scale related metabolic models made work together with flux balance analysis algorithms would allow *in silico* simulation of organism's growth under certain metabolic restrictions that are based on availability of substrates in the medium. The growth simulations obtained can then be utilized to better optimize medium compositions that better fitted the necessity of the organism (Alkema et al., 2016).

In case of fermentation processes, bioinformatics techniques are helpful to analyse the interaction of bacteriophages and microbes and provide a way for food scientists to analyse the metabolic necessity of the microbial consortia that is observed during fermentation. This investigation could lead a way in future to gain information related to stability in fermentation. Initially, it could be attained by carrying out experiments with the help of synthetic microbial community. The design based on this community of microbes is currently being explored and cross-kingdom interactions are being investigated progressively (Bernstein and Carlson, 2014; Mansour et al., 2009). In a previous study by Wolfe et al. (2014) cheese rind-based bacterial communities were made with the help of available omics data, follow-up experiments, and knowledge on fermentation, which eventually demonstrated the capacity of predicting features of complex fermentations. Based on these existing literature, this field of microbial informatics technology could be considered as the future of bioinformatics in food safety.

Applications of Bioinformatics in Food Science

Detection of Bioactive Peptides

Food-derived peptides possess functional activities that are far beyond their elementary role in nutrition. Recently, there is a growing interest in the areas of both academic and commercial firms in the utilization of foods that are augmented with bioactive peptides which could promote health benefits and prevent diseased conditions. Protein precursors synthesized by fermentation of bacteria or activity of gastrointestinal enzymes or additional proteolytic enzymes release bioactive peptides that are based on food sources. Though many conventional methods have been found to identify bioactive peptides synthesized by this way, it has now been very evident that application of bioinformatics can be considered as an innovative strategy for identification of novel bioactive peptides (Khaldi, 2012).

The approach utilizes data found in database like BIOPEP (Waidha et al., 2016). The protein sequences from SwissProt (https://www.expasy.org/resources/uniprotkb-swiss-prot), UniProtKB and TreMBL (http://www.bioinfo.pte.hu/more/TrEMBL.htm) databases can be retrieved to determine how frequently bioactive peptide are seen in primary structures of food proteins and can be evaluated by a/N ratio where "a" denotes the number of peptides which exhibit bioactivity of a type of peptide found in sequence and "N" denotes the total number of amino acid residues found within the proteins. Computational tools based software were used to create profiles for *in silico* peptides from the results of proteolytic specificities of various enzymes (Chibuike, 2014). Some of the currently available proteolytic analysis tools include BIOPEP, ExPASY (https://www.expasy.org/) and PoPS (https://openebench.bsc.es/tool/pops-protease). These processes could reduce time and can prompt the identification of new prototype of known bioactive peptides (Waidha et al., 2016).

Food Quality and Safety

An enormous growth is witnessed on applying bioinformatics technologies in the domain of food quality and safety. Genome sequencing projects are currently focused on innovative ways for determining the foodborne pathogens which are the causative organism for various food borne illnesses (Brul et al., 2006). Molecular markers would aid in identification of pathogenic and spoilage bacteria and could predict preservation stress resistance in future. Bioinformaticians have generated various tools which was released by FDA (Food and Drug Administration) to detect and identify these foodborne pathogens in an efficient way. Using these tools, the molecular characterization of food borne pathogens can be studied by the technique of microarrays (Fang et al., 2010).

In addition, the next generation sequencing technologies has now emerged as a novel method to bioinformatically investigate the causative agent responsible for illness caused due to foodborne pathogens (Lienau et al., 2011). Other computational techniques related to this field is the application of neural networks to predict the growth of microbes in a given food source (García-Gimeno, 2002). The research concerned to food quality, still progresses in the field of computer vision that would allow automatic estimation of various food properties in future (Mery et al., 2012).

Food Taste and Flavour

Different taste receptors for tastes including sour, bitter, umami, sweet, and salt were identified. With the help of these taste receptors the next generation taste modifiers were developed for foods. Using the techniques of molecular modelling and simulations and based on known structures of these receptors, innovative computational software and algorithms were developed. These simulations resulted in creation of intense tasting constituents as food additives and provided a better understanding on the basis of taste antagonism, persistence, and complementation. For instance, the homology between receptors of sweet taste and brain glutamate and sour taste sensors in mammals were determined using the bioinformatics approach of sequence similarity algorithms (Talevi et al., 2012).

A major compound in flavour formation related to dairy products is lactic acid. The flavour forming capacity of lactic acid bacteria was identified by analysing the genetic sequences of lactic acid bacteria. The flavouring properties of various food products are due to the interaction of different compounds and not restricted to single compounds alone. Computational technologies are considered as an effective platform for linking various flavour compounds based on preferences and needs of the consumer for development of new products (Liu et al., 2009).

Detection of Allergen

Bioinformatics are widely used in analysis of food allergens. Most of the allergenic constituents are similar in their sequence and structure. Hence, techniques of structural bioinformatics and homology modelling could detect cross reactivity of proteins and probable allergens in food products. For evaluating the allergenicity of genetically modified foods, World Health Organization (WHO) has incorporated sequences similarity search as a part of the guidelines in detection of allergens (Jenkins et al., 2005). Few of the existing databases related to food allergens detection include Informall FARRP Allergen database (https://farrp.unl.edu/informallbig8), AllerMatch (http://www.allermatch.org/) and SDAP (https://fermi.utmb.edu/) that have greatly influenced the applicability of bioinformatics in this field (Mari et al., 2006; Fiers et al., 2004).

Other bioinformatics area that were of heightened interests recently are predicting allergens based on machine learning tools namely support vector machines (Saha & Raghava, 2006), amino acid properties (Dimitrov et al., 2013) and common sequence based motifs (Ivanciuc et al., 2009). In such a case, Lin et al. (2012) has implemented machine learning method to predict peanut allergy through development of allergen diagnostics.

Pre and Probiotics

There is an increase in the vending of pre and probiotic foods, as the multiple microbial flora found in the gastrointestinal tract of humans are required for prevention of numerous diseases and healthy intestinal function. With the onset of genomics era, the application of bioinformatics in the analysis of pre and probiotics mechanistically has greatly propelled (Pridmore et al., 2004). Recent bioinformatic applications in combination with molecular and mass spectroscopy technologies have influenced the study of gut microbiota with an insight knowledge on the metabolites present in the host obtained due to interaction (Rath et al., 2012). Researchers could now select the probiotics depending on subjects age by applying advanced high throughput sequencing technologies (Dominguez-Bello et al., 2011).

Utilization of bioinformatics in the field of prebiotics are of great interest as it is associated with various health benefits, such as obesity, inhibition of cancer, and reduction of cholesterol. Apart from this, only a minor bioinformatic research is merely committed to the study of prebiotics and maximum studies are focused on pre and probiotics together (Collison et al., 2012). A study by Everard et al. (2011) reported bioinformatics technique namely phylogenetic microarrays which has been shown to investigate the gut microbiota in diabetic and obese mice after prebiotic administration.

Construction of Metabolic Pathway

Microbial metabolism has been employed in fermentation of foods. By understanding the mechanism of microbial metabolism, the efficacy of fermentation of foods could be greatly enhanced. With the application of desired microbes, it can be converted to substrate (especially carbohydrates) to organic compounds which contributes to structure, flavour, stability, safety, and texture of food products. Current bioinformatic and genomic methods have expanded the available tools to mechanistically decipher and prevent metabolism of microbes. The databases related to genomic sequences of bacteria and information on metabolic networks from literature and stochiometric models describes biological processes and interlink phenotype and genotype.

With an increasing knowledge in this research area, more metabolic reconstruction models could be considered to study the cell's response to external stimuli dynamically (Waidha et al., 2016). As a holistic approach, biological processes utilize genomic information to recreate complete metabolic pathways. Using this approach, expression level and annotated genes from new organism could be matched with the known metabolic pathways stored in already existing databases (Rani and Gomathi, 2017).

Crop Production

Genetically modified crops developed by the approach of genetic research in plants have enormously increased the production rate in food industries. The foremost objective of this technology is to improve the efficiency and quality of raw materials that are supplied for food supply so as to provide a safer food and lower the cost. Identification of biosynthetic genes in plants are possible because of genome sequencing projects. Thus genome research could directly be incorporated in order to enhance the efficacy in breeding of plants for improvement (Waidha et al., 2016).

One of the most commonly used raw material in every food product containing chocolate is *Theobroma cacao i.e.* Cocoa. For the past years, it was difficult to identify a higher quality seed with best flavor effectively. With the advancement of DNA fingerprinting, on screening of various plants to detect RFPL markers that can identify connection between breeds, more than 300 markers breeding programs has been extensively improved for identification. Additionally, techniques like expressed sequence tags (EST) genetic comparisons and sequences to other identified plants largely depends on bioinformatics techniques which could further increase the possibility to choose the traits of interest during the early stage of development in plants (Pridmore et al., 2000).

Food Processing

Application of bioinformatics tools and techniques in food processing could optimize the compositional and quantitative parameters of certain traditional operations. In general, to enhance the safety and storage ability, food products are processed effectively. Wide array of errors occur due to lack of knowledge about the structural complexity and composition of biological materials and as a result, there is a variation in living organisms. These living organisms were used as food process input materials and responses are considered as processing parameters. With an extensive knowledge on utilization of microorganisms in the field of food processing and acquired bioinformatics tools there might be a decrease in the number of errors. The technique of food processing involved reconstruction of biomaterials with the aid of high energy input and combining biological intellect with the knowledge of biomaterials.

Reconstruction used simple macrostructures from organisms to form stable foods. But, during the elimination of hazardous features of some molecules, the most significant immanent biological features of the final product are abandoned. In order to avoid this circumstance, an in-depth knowledge on bioinformatics tools could demonstrate the most important features responsible for complex structures of biomolecules and its functions to yield a precise and stable analysis of biomaterial properties of molecules in a fusion of complexes.

For instance, possible interactions among the 6000 yeast proteins referred to as protein-protein synergy map of yeast have been completed. Thus the research carried out in this manner by exploitation of constructional features of living organisms with the help of bioinformatics will eventually show its effects on food processes and food-products in future (Waidha et al., 2016).

Conclusion

In conclusion, food science is becoming a data augmented subject compared to other branch of sciences. The field of bioinformatics have been increasingly gaining popularity for diverse applications in the area of food science and offers vast potential scope for the development of new technologies to implement them in food based industries. For large amount of data obtained through high throughput "omics" technologies implying bioinformatics would be suitable to achieve its purpose in the arena of food science. These data generated from outlined projects could make bioinformatics a well-established and important area in food industries. In order to handle enormous amount of data efficiently that are developed from different projects, bioinformatics should be utilized effectively in food industries. Various attempts are initiated via bioinformatics tools to develop genetically engineered transgenic plants in order to obtain beneficial and nutritious food materials for the public and may eventually serve as a breakthrough in the area of food science. Through this breakthrough approaches, all food based industries will be more benefitted accurately and efficiently. In future, more extensive technologies pertained to bioinformatics in food industries such as designing of various models based on knowledge of genes and their functions could be implemented. Though this developmental approach is at rise, the full commitment of this approach is yet to be explored.

References

Abee, T., Van Schaik, W., and Siezen, R.J. (2004). Impact of genomics on microbial food safety, Trends in Biotechnology, 22, 653-660.

Alkema, W., Boekhorst, J., Wels, M., and Hijum, S.A.F.T. (2016). Microbial bioinformatics for food safety and production, Briefings in bioinformatics, 17(2), 283-292.

Allard, M.W., Bell, R., Ferreira, C.M., Gonzalez-Escalona, N., Hoffmann, M., Muruvanda, T., Ottesen, A., Ramachandran, P., Reed, E., Sharma, S., Stevens, E., Timme, R., Zheng, J., and Brown, E.W. (2017). Genomics of foodborne pathogens for microbial food safety, Current Opinion in Biotechnology, 49, 224-229.

Alvarez-Rivera, G., Toro-Uribe, S., Ibanez, E., Cifuentes, A., and Bueno, M. (2018). Omics technology: foodomics, In: Reference Module in Chemistry, Molecular Sciences and Chemical Engineering.

Barretto, C., Rincón, C., Portmann, A.C., and Ngom-Bru, C. (2021). Whole Genome Sequencing Applied to Pathogen Source Tracking in Food Industry: Key Considerations for Robust Bioinformatics Data Analysis and Reliable Results Interpretation, Genes, 12, 275.

Bendixen, E., Danielsen, M., Hollung, K., Gianazza, E., and Miller, I. (2011). Farm animal proteomics- a review, Journal of Proteomics, 74(3), 282-293.

Bentley, S.D., and Parkhill, J. (2004). Comparative genomic structure of prokaryotes, Annual Review of Genetics, 38, 771-792.

Bernstein, H.C., and Carlson, R.P. (2014). Design, construction, and characterization methodologies for synthetic microbial consortia, Methods in Molecular Biology, 1151, 49-68.

Brul, S., Schuren, F., Montijn, R., Keijser, B.J.F., Van Der Spek H., et al. (2006) The impact of functional genomics on microbiological food quality and safety, International Journal of Food Microbiology, 112, 195-199.

Chibuike C. Udenigwe. (2014). Bioinformatics approaches prospects and challenges of food bioactive peptide research, Trends in Food Science & Technology, 36(2), 137-143.

Chun, J., Grim, C.J., Hasan, N.A., Lee, J.H., Choi, S.Y., Haley, B.J., et al. (2009). Comparative genomics reveals mechanism for short-term and long-term clonal transitions in pandemic Vibrio cholerae, Proceedings of the National Academy of Sciences, 106, 15442-15447.

Cifuentes, A. (2017). Foodomics, foodome and modern food analysis, TrAC Trends in Analytical Chemistry, 96, 1.

Collison, M., Hirt, R.P., Wipat, A., Nakjang, S., Sanseau, P., and Brown, J.R. (2012). Data mining the human gut microbiota for therapeutic targets, Briefings in Bioinformatics, 13, 751-768.

Dimitrov, I., Flower, D., and Doytchinova, I. (2013). AllerTOP a server for *in silico* prediction of allergens, BMC Bioinformatics, 14, S4.

Dominguez-Bello, M.G., Blaser, M.J., Ley, R.E., and Knight, R. (2011). Development of the human gastrointestinal microbiota and insights from high-throughput sequencing, Gastroenterology, 140, 1713-1719.

Everard, A., Lazarevic, V., Derrien, M., Girard, M., Muccioli, G.G., Neyrinck, A.M., et al. (2011) Responses of gut microbiota and glucose and lipid metabolism to prebiotics in genetic obese and diet-induced leptin-resistant mice, Diabetes, 60, 2775-2786.

Fang, H., Xu, J., Ding, D., Jackson, S.A., Patel, I.R., et al. (2010). An FDA bioinformatics tool for microbial genomics research on molecular characterization of bacterial foodborne pathogens using microarrays, BMC Bioinformatics, 11, S4.

Fang, H.Y., Chen, S.B., Guo, D.J., Pan, S.Y., and Yu, Z.L. (2011). Proteomic identification of differentially expressed proteins in curcumin-treated MCF-7 cells, Phytomedicine, 18(8-9), 697-703.

Fei, Z., Tang, X., Alba, R.M., White, J.A., Ronning, C.M., Martin, G.B., et al. (2004). Comprehensive EST analysis of tomato and comparative genomics of fruit ripening, The Plant Journal, 40, 47-59.

Fiers, M.W., Kleter, G.A., Nijland, H., Peijnenburg, A.A., Nap, J.P., et al. (2004). Allermatch™, a webtool for the prediction of potential allergenicity according to current FAO/WHO Codex alimentarius guidelines, BMC Bioinformatics, 5, 133.

Garcıa-Gimeno, R.M.A., Hervas-Martinez, C., and De Silonizc, M.I. (2002). Improving artificial neural networks with a pruning methodology and genetic algorithms for their application in microbial growth prediction in food, International Journal of Food Microbiology, 72, 19-30.

Gry, J., Black, L., Eriksen, F.D., Pilegaard, K., Plumb, J., et al. (2007). EuroFIRBASIS–a combined composition and biological activity database for bioactive compounds in plant-based foods, Trends in Food Science and Technology, 18, 434-444.

Healy, M.J., Tong, W., Ostroff, S., Eichler, H.G., Patak, A., Neuspiel, M., Deluyker, H., and Slikker Jr.,W. (2016). Regulatory bioinformatics for food and drug safety, Regulatory Toxicology and Pharmacology.

Hoffmann, S., Devleesschauwer, B., Aspinall, W., Cooke, R., Corrigan, T., Havelaar, A., Angulo, F., Gibb, H., Kirk, M., Lake, R., and Speybroeck, N. (2017). Attribution of global foodborne disease to specific foods: findings from a World Health Organization structured expert elicitation, PLoS One, 12(9), e0183641.

Holton, T.A., Vijayakumar, V., and Khaldi, N. (2012). Bioinformatics: current perspectives and future directions for food and nutritional research facilitated by a Food-Wiki database, Trends in Food Science and Technology, 34(1), 5-17.

Ivanciuc, O., Garcia, T., Torres, M., Schein, C.H., and Braun, W. (2009). Characteristic motifs for families of allergenic proteins, Molecular Immunology, 46, 559-568.

Jackson, B.R., Tarr, C., Strain, E., Jackson, K.A., Conrad, A., Carleton, H., Katz, L.S., Stroika, S., Gould, L.H., Mody, R.K., Silk, B.J., Beal, J., Chen, Y., Timme, R., Doyle, M., Fields, A., Wise, M., Tillman, G., Defibaugh-Chavez, S., Kucerova, Z., Sabol, A., Roache, K., Trees, E., Simmons, M., Wasilenko, J., Kubota, K., Pouseele, H., Klimke, W., Besser, J., Brown, E., Allard, M., and Gerner-Smidt, P. (2016). Implementation of nationwide real-time whole- genome sequencing to enhance Listeriosis outbreak detection and investigation, Clinical Infectious Diseases, 63, 380-386.

Jenkins, J.A., Griffiths-Jones, S., Shewry, P.R., Breiteneder, H., and Mills, E.C. (2005). Structural relatedness of plant food allergens with specific reference to cross-reactive allergens: an in silico analysis, Journal of Allergy and Clinical Immunology, 115, 163-170.

Khaldi, N. (2012). Bioinformatics approaches for identifying new therapeutic bioactive peptides in food, Functional Foods in Health and Disease, 2, 325-338.

Khaldi, N., and Wolfe, K.H. (2011). Evolutionary origins of the fumonisin secondary metabolite gene cluster in *Fusarium verticillioides* and *Aspergillus niger*, International Journal of Evolutionary Biology, 423821.

Khaldi, N., Seifuddin, F.T., Turner, G., Haft, D., Nierman, W.C., Wolfe, K.H., et al. (2010). SMURF: genomic mapping of fungal secondary metabolite clusters, Fungal Genetics and Biology, 47, 736-741.

Kumar, A., and Chordia, N. (2017). Bioinformatics approaches in food sciences, Journal of Food: Microbiology, Safety and Hygiene, 2, 2.

Lemay, D.G., Zivkovic, A.M., and German, J.B. (2007). Building the bridges to bioinformatics in nutrition research, The American Journal of Clinical Nutrition, 86, 1261-1269.

Lienau, E.K., Strain, E., Wang, C., Zheng, J., Ottesen, A.R., Keys, C.E., et al. (2011). Identification of a salmonellosis outbreak by means of molecular sequencing, New England Journal of Medicine, 364, 981-982.

Lin, J., Bruni, F.M., Fu, Z., Maloney, J., Bardina, L., Boner, A.L., et al. (2012). A bioinformatics approach to identify patients with symptomatic peanut allergy using peptide microarray immunoassay, Journal of Allergy and Clinical Immunology, 129, 1321-1328.

Liu, M., Nauta, A., Francke, C., and Siezen, R.J. (2008). Comparative genomics of enzymes in flavor-forming pathways from amino acids in lactic acid bacteria, Applied and Environmental Microbiology, 74, 4590-4600.

Ma, L.J., Van Der Does, H.C., Borkovich, K.A., Coleman, J.J., Daboussi, M.J., Di Pietro, A., et al. (2010). Comparative genomics reveals mobile pathogenicity chromosomes in Fusarium, Nature, 464, 367-373.

Mansour, S., Bailly, J., Landaud, S., et al. (2009). Investigation of associations of *Yarrowia lipolytica*, *Staphylococcus xylosus*, and *Lactococcus lactis* in culture as a first step in microbial interaction analysis. Applied and Environmental Microbiology, 75, 6422-6430.

Mari, A., Scala, E., Palazzo, P., Ridolfi, S., Zennaro, D., et al. (2006). Bioinformatics applied to allergy: allergen databases, from collecting sequence information to data integration. The Allergome platform as a model, Cellular immune, 244, 97-100.

Mayer, K.F.X., Martis, M., Hedley, P.E., Simkova, H., Liu, H., Morris, J.A., et al. (2011). Unlocking the barley genome by chromosomal and comparative genomics, The Plant Cell Online, 23, 1249-1263.

Mehta, A., and Hasija, Y. (2018). Bioinformatics Approaches for Genomics and Post Genomics Applications of Anticancer Plants in Anticancer Plants: Mechanisms and Molecular Interactions, Springer, Singapore, 283-317.

Merchant, A., and Dehghan, M. (2006). Food composition database development for between country comparisons, Nutrition Journal, 5, 2.

Mery, D., Pedreschi, F., and Soto, A. (2012). Automated design of a computer vision system for visual food quality evaluation, Food and Bioprocess Technology, 1-16.

Mochida, K., and Shinozaki, K. (2010). Genomics and bioinformatics resources for crop improvement, Plant and Cell Physiology, 51, 497-523.

Ogura, Y., Ooka, T., Iguchi, A., Toh, H., Asadulghani, M., Oshima, K., et al. (2009). Comparative genomics reveal the mechanism of the parallel evolution of O157 and non-O157 enterohemorrhagic *Escherichia coli*, Proceedings of the National Academy of Sciences, 106, 17939-17944.

Omelchenko, M., Wolf, Y., Gaidamakova, E., Matrosova, V., Vasilenko, A., Zhai, M., et al. (2005). Comparative genomics of *Thermus thermophilus* and *Deinococcus radiodurans*: divergent routes of adaptation to thermophily and radiation resistance, BMC Evolutionary Biology, 5, 57.

Pennington, J.A.T., Stumbo, P.J., Murphy, S.P., Mcnutt, S.W., Eldridge, A.L., Mccabe-Sellers, B.J., et al. (2007). Food composition data: the foundation of dietetic practice and research, Journal of the American Dietetic Association, 107, 2105-2113.

Prakash, B., Singh, P.P., Kumar, A., and Gupta, V. (2020). Prospects of omics technologies and bioinformatics approaches in food science, Functional and preservative properties of phytochemicals, Elsevier Co. publishers.

Pridmore, R.D., Berger, B., Desiere, F., Vilanova, D., Barretto, C., Pittet, A.C., et al. (2004). The genome sequence of the probiotic intestinal bacterium Lactobacillus johnsonii NCC 533, Proceedings of the National Academy of Sciences of the United States of America, 101, 2512-2517.

Pridmore, R.D., Crouzillat, D., Walker, C., Foley, S., Zink, R., Zwahlen, M.C., Brüssow, H., Pétiard, V., and Mollet, B. (2000). Genomics molecular genetics and the food industry, Journal of Biotechnology, 78(3), 251-258.

Qian, W., Jia, Y., Ren, S.X., He, Y.Q., Feng, J.X., Lu, L.F., et al. (2005). Comparative and functional genomic analyses of the pathogenicity of phytopathogen *Xanthomonas campestris* pv. campestris, Genome Research, 15, 757-767.

Rani, K., and Gomathi, K. (2017). Application of Bioinformatics in Food Quality Control, International Journal of Current Research and Modern Education, Special Issue: 43-46.

Rath, C., Alexandrov, T., Higginbotton, S., Song, J., Milla, M., Fischbach, M., et al. (2012). Molecular analysis of model gut microbiotas by imaging mass spectrometry and nano-desorption electrospray ionization reveals dietary metabolite transformations, Analytical Chemistry, 84, 9259-9267.

Saha, S.; and Raghava, G.P.S. (2006). AlgPred: prediction of allergenic proteins and mapping of IgE epitopes, Nucleic Acids Research, 34, 202-209.

Taboada E.N., Graham M.R., Carriço J.A., and Van Domselaar G. (2017) Food Safety in the Age of Next Generation Sequencing, Bioinformatics, and Open Data Access, Frontiers in Microbiology, 8, 909.

Talevi, A., Enrique, A.V., and Bruno-Blanch, L.E. (2012). Anticonvulsant activity of artificial sweeteners: a structural link between sweet-taste receptor T1R3 and brain glutamate receptors, Bioorganic & Medicinal Chemistry Letters, 22, 4072-4074.

Tong, W., Ostroff, S., Blais, B., Silva, P., Dubuc, M., Healy, M., and Slikker, W. (2015). Genomics in the land of regulatory science, Regulatory Toxicology and Pharmacology, 72(1), 102-106.

Tu, M., Cheng, S., Lu, W., and Du, M. (2018). Advancement and prospects of bioinformatics analysis for studying bioactive peptides from food-derived protein: sequence, structure, and functions, TrAC Trends in Analytical Chemistry, 105, 7-17.

Valdés, A., Cifuentes, A., and Leon, C. (2017). Foodomics evaluation of bioactive compounds in foods, TrAC Trends in Analytical Chemistry, 96, 2-13.

Waidha, K.M., Jabalia, N., Singh, D., Jha, A., and Kaur, R. (2016). Bioinformatics approaches in food industry: An overview, National conference on recent trends in biomedical engineering, cancer biology, bioinformatics and applied biotechnology.

Wohlbach, D.J., Kuo, A., Sato, T.K., Potts, K.M., Salamov, A.A., Labutti, K.M., et al. (2011). Comparative genomics of xylose-fermenting fungi for enhanced biofuel production, Proceedings of the National Academy of Sciences, 108, 13212-13217.

Wolfe, B.E., Button, J.E., Santarelli, M., et al. (2014). Cheese rind communities provide tractable systems for *in situ* and *in vitro* studies of microbial diversity, Cell, 158, 422-33.

World Health Organization Food and Agriculture Organization of the United Nations. (2019). Applications of Whole Genome Sequencing in Food Safety Management Food and Agriculture Organization of the United Nations in Collaboration with the Rome, 2016. Available at: http://www.fao.org/3/a-i5619e.pdf.

Xu, W. (2018). Application of bioinformatics technology in food processing and food safety assessment, International conference on Zoology, Botany, and Ecology.

Index

P

R

S

T

V

W

[illegible]

Shankar, 101, 103 [illegible]

[illegible] 134, 14[illegible]

[illegible] 202, 231 [illegible]

[illegible]

[illegible]

[illegible]